Chlorinated Organic Micropollutants

ISSUES IN ENVIRONMENTAL SCIENCE AND TECHNOLOGY

EDITORS:

R. E. Hester, University of York, UK
R. M. Harrison, University of Birmingham, UK

EDITORIAL ADVISORY BOARD:

A. K. Barbour, Specialist in Environmental Science and Regulation, UK, **N. A. Burdett,** National Power PLC, UK, **J. Cairns, Jr.,** Virginia Polytechnic Institute and State University, USA, **P. A. Chave,** Pollution Control Consultant, UK, **P. Crutzen,** Max-Planck-Institut für Chemie, Germany, **P. Doyle,** Zeneca Group PLC, UK, **Sir Hugh Fish,** Consultant, UK, **M. J. Gittins,** Leeds City Council, UK, **J. E. Harries,** Imperial College, London, UK, **P. K. Hopke,** Clarkson University, USA, **Sir John Houghton,** Royal Commission on Environmental Pollution, UK, **N. J. King,** Environmental Consultant, UK, **S. Matsui,** Kyoto University, Japan, **D. H. Slater,** Environment Agency, UK, **T. G. Spiro,** Princeton University, USA, **D. Taylor,** Zeneca Limited, UK, **Sir Frederick Warner,** SCOPE Office, UK.

TITLES IN THE SERIES:

1 Mining and its Environmental Impact
2 Waste Incineration and the Environment
3 Waste Treatment and Disposal
4 Volatile Organic Compounds in the Atmosphere
5 Agricultural Chemicals and the Environment
6 Chlorinated Organic Micropollutants

FORTHCOMING:

7 Contaminated Land and Its Reclamation
8 Air Quality Management

How to obtain future titles on publication

A subscription is available for this series. This will bring delivery of each new volume immediately upon publication. For further information, please write to:

The Royal Society of Chemistry
Turpin Distribution Services Limited
Blackhorse Road
Letchworth
Herts SG6 1HN, UK

Telephone: +44 (0) 1462 672555
Fax: +44 (0) 1462 480947

ISSUES IN ENVIRONMENTAL SCIENCE
AND TECHNOLOGY

EDITORS: R. E. HESTER AND R. M. HARRISON

6

Chlorinated Organic Micropollutants

THE ROYAL
SOCIETY OF
CHEMISTRY
Information
Services

ISBN 0-85404-225-3
ISSN 1350-7583

A catalogue record for this book is available from the British Library

Published by The Royal Society of Chemistry, Thomas Graham House,
Science Park, Milton Road, Cambridge CB4 4WF, UK

Typeset in Great Britain by Vision Typesetting, Manchester
Printed and bound in Great Britain by Bookcraft (Bath) Ltd

Preface

In this volume of Issues we address the sources, environmental cycles, uptake, consequences and control of many of the more important chlorinated organic micropollutants. Under this heading we have included a range of semi-volatile persistent compounds, notably polychlorinated biphenyls (PCBs), polychlorinated dibenzo-*p*-dioxins (PCDDs) and polychlorinated dibenzofurans (PCDFs) as well as a number of chlorinated pesticides. We have not sought to include volatile species such as CFCs which cause environmental problems of an entirely different nature. The compounds included in this volume cause no threat to the stratospheric ozone layer, but have given widespread cause for concern in relation to their environmental persistence and high toxicity, and their potential for adverse effects on humans and wildlife.

Despite the fact that PCBs and some of the chlorinated pesticides are no longer manufactured, they remain relatively abundant in the environment because of their low reactivity. A further consequence of this chemical inertness and their lipid solubility is a tendency to concentrate within food chains and hence present the greatest level of risk to those at the top of the food chain. In the case of PCDDs and PCDFs, there is evidence of natural production from sources such as forest fires, but this appears to be modest in magnitude, and current environmental burdens result largely from human activity. Prior to the Seveso incident few had heard of these compounds, whereas nowadays 'dioxins' are regarded as major environmental hazards by the general public, who derive their opinions largely from poorly informed press coverage, itself often fuelled by incomplete and sometimes inaccurate information put forward by pressure groups, but reflecting also some genuine disagreements within the scientific community over the risks posed by these compounds.

Rational decision making over chlorinated organic micropollutants in the environment must be based upon sound science. This volume draws upon the expertise of some of the most distinguished workers in this field to review current knowledge of the sources, environmental concentrations and pathways, human toxicity and ecotoxicology, and control methods for these groups of compounds. In the first article, Harrad addresses some of the problems of quantification inherent in understanding the environmental inventories and budgets of PCBs, PCDDs and PCDFs. The source inventory approach is extended by Travis and Nixon in the second article to evaluate sources of human exposures to PCDDs and PCDFs. Such exposure to these compounds depends crucially upon biological uptake and transfers through the food chain, for example, from atmospheric emissions into pasture grass and thus into cows' milk; McLachlan provides a detailed insight into the processes involved and their relative efficiencies. Despite their known persistence in the environment, PCBs, dioxins and furans are both decomposed within, and removed from, the atmosphere by scavenging processes, and Atkinson reviews knowledge of the processes involved. The next two articles by Safe and de Voogt describe, respectively, the human toxicology and ecotoxicology of exposure to chlorinated organic micropollutants. These put into clear context the consequences of the exposures estimated through the early chapters. Turnbull then focuses on the usage, environmental cycles and concentrations of chlorinated pesticides, showing that even compounds which are subject to extensive bans upon production and use are still cycling within the environment. One of the ecosystems which has suffered the greatest pollution from PCBs is the North American Great Lakes system. This has been the subject of intensive scientific investigation which is described as a case study by Swackhamer. Historically, one of the most important source categories for 'dioxin' emissions has been the combustion of waste in incinerators. Since the 'dioxin' problem became evident, far more stringent controls have been applied in developed countries to incinerator emissions and much has been learnt about the optimal techniques for controlling PCDD and PCDF formation and emission. In the final article, Eduljee and Cains describe the operating procedures and control technologies available for minimizing such emissions.

We believe that this volume gives a unique and valuable compilation of information on an extremely important group of environmental pollutants. It is fully up-to-date and should provide a comprehensive overview of this topical subject useful for some years to come.

Ronald E. Hester
Roy M. Harrison

Contents

Editors

Ronald E. Hester, BSc, DSc(London), PhD(Cornell), FRSC, CChem

Ronald E. Hester is Professor of Chemistry in the University of York. He was for short periods a research fellow in Cambridge and an assistant professor at Cornell before being appointed to a lectureship in chemistry in York in 1965. He has been a full professor in York since 1983. His more than 300 publications are mainly in the area of vibrational spectroscopy, latterly focusing on time-resolved studies of photoreaction intermediates and on biomolecular systems in solution. He is active in environmental chemistry and is a founder member and former chairman of the Environment Group of The Royal Society of Chemistry and editor of 'Industry and the Environment in Perspective' (RSC, 1983) and 'Understanding Our Environment' (RSC, 1986). As a member of the Council of the UK Science and Engineering Research Council and several of its sub-committees, panels and boards, he has been heavily involved in national science policy and administration. He was, from 1991–93, a member of the UK Department of the Environment Advisory Committee on Hazardous Substances and is currently a member of the Publications and Information Board of The Royal Society of Chemistry.

Roy M. Harrison, BSc, PhD, DSc (Birmingham), FRSC, CChem, FRMetS, FRSH

Roy M. Harrison is Queen Elizabeth II Birmingham Centenary Professor of Environmental Health in the University of Birmingham. He was previously Lecturer in Environmental Sciences at the University of Lancaster and Reader and Director of the Institute of Aerosol Science at the University of Essex. His more than 250 publications are mainly in the field of environmental chemistry, although his current work includes studies of human health impacts of atmospheric pollutants as well as research into the chemistry of pollution phenomena. He is a former member and past Chairman of the Environment Group of The Royal Society of Chemistry for whom he has edited 'Pollution: Causes, Effects and Control' (RSC, 1983; Third Edition, 1996) and 'Understanding our Environment: An Introduction to Environmental Chemistry and Pollution' (RSC, Second Edition, 1992). He has a close interest in scientific and policy aspects of air pollution, currently being Chairman of the Department of Environment Quality of Urban Air Review Group as well as a member of the DoE Expert Panel on Air Quality Standards and Photochemical Oxidants Review Group and the Department of Health Committee on the Medical Effects of Air Pollutants.

Contributors

R. Atkinson, *University of California, Statewide Air Pollution Research Center, Riverside, CA 92521, USA*

P. Cains, *Environmental and Process Engineering, AEA Technology, 404 Harwell, Didcot, Oxon OX11 0RA, UK*

P. de Voogt, *Amsterdam Research Institute for Substances in Ecosystems, Department of Environmental and Toxicological Chemistry, University of Amsterdam, Nieuwe Achtergracht 166, 1018 WV Amsterdam, Netherlands*

G. H. Eduljee, *Environmental Resources Management, Eaton House, Wallbrook Court, North Hinksey Lane, Oxford OX2 0QS, UK*

S. J. Harrad, *School of Chemistry, Institute of Public and Environmental Health, The University of Birmingham, Edgbaston, Birmingham B15 2TT, UK*

M. McLachlan, *Ecological Chemistry and Geochemistry, University of Bayreuth, D-95400 Bayreuth, Germany*

A. G. Nixon, *University of Tennessee, Knoxville, TN 37996, USA*

S. Safe, *Department of Veterinary Physiology and Pharmacology, Texas A & M University, College Station, TX 77843-4466, USA*

D. L. Swackhamer, *Environmental and Occupational Health, School of Public Health, Box 807 UMHC, 420 Delaware St SE, University of Minnesota, Minneapolis, MN 55455, USA*

C. C. Travis, *Health Sciences Research Division, Oak Ridge National Laboratory, 1060 Commerce Park Drive, Oak Ridge, TN 37830, USA*

A. B. Turnbull, *School of Chemistry, Institute of Environmental and Public Health, The University of Birmingham, Edgbaston, Birmingham B15 2TT, UK*

Sources and Fates of Polychlorinated Dibenzo-*p*-dioxins, Dibenzofurans and Biphenyls: The Budget and Source Inventory Approach

STUART J. HARRAD

1 Introduction

Polychlorinated dibenzo-*p*-dioxins (PCDDs), polychlorinated dibenzofurans (PCDFs) and polychlorinated biphenyls (PCBs) have attracted considerable attention in recent decades, owing to concern over their potential adverse effects in humans and wildlife, which are compounded by their ubiquitous environmental presence and resistance to degradation. Amongst the 75 possible PCDDs, 135 PCDFs and 209 PCBs, there exists wide variation in physicochemical properties, bioaccumulative tendencies and toxicity. Figures 1 and 2 illustrate the basic structures and nomenclature of both PCDDs, PCDFs – collectively referred to as PCDD/Fs – and PCBs.

This chapter reviews our knowledge of several key issues pertaining to the environmental presence of these compounds. Constructing source inventories for

Figure 1 Basic structures and nomenclature of PCDD/Fs

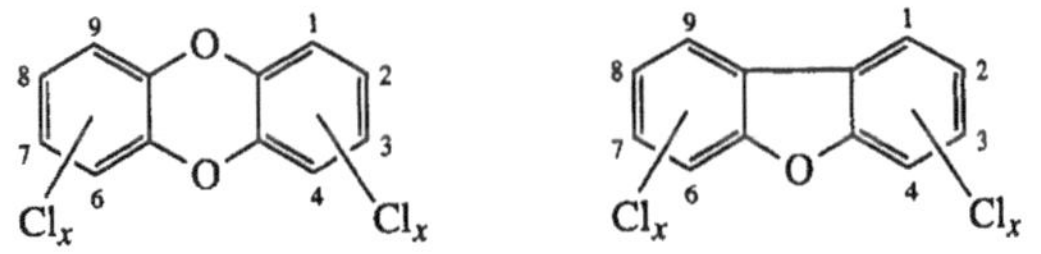

A PCDD $x = 0 - 4$ A PCDF

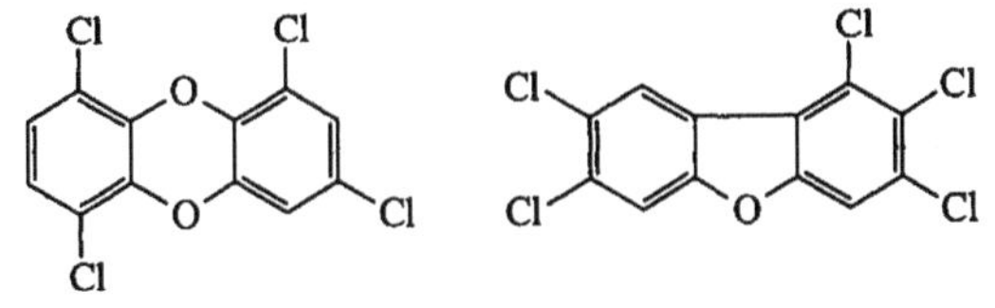

1,3,6,9-TCDD   1,2,3,7,8-PeCDF

Figure 2 Basic structures and nomenclature of PCBs

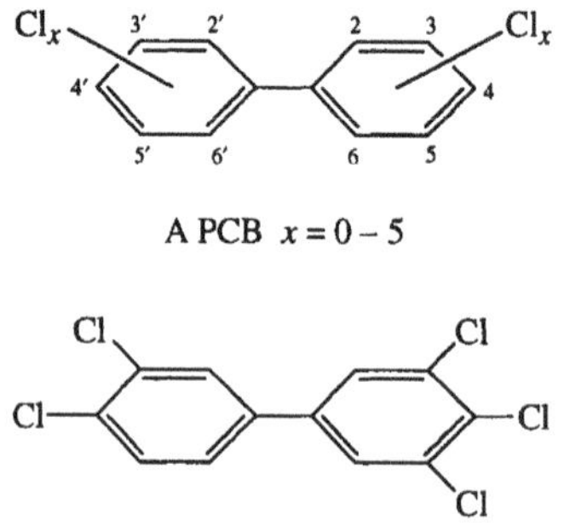

A PCB $x = 0 - 5$

3,3′,4,4′,5-Pentachlorobiphenyl – PCB 126

a group of chemical pollutants permits the targeting of specific sources in order to reduce environmental emissions and hence human exposure, whilst the establishment of environmental budgets facilitates the identification of major reservoirs, and quantification of the extent to which a given pollutant has been released into the environment and been subsequently 'lost' *via* either biodegradation or environmental transport.

2 Physicochemical Properties and Environmental Levels

PCDD/Fs and PCBs possess low vapour pressures and water solubilities, along with high octan-1-ol/water partition coefficients (K_{ow} values), which are listed for selected congeners in Table 1. When the long biological lifetimes of these chemicals are taken into account (human half-lives of up to 27.5 years have been reported for some PCBs[1]), it is unsurprising that PCDD/Fs and PCBs display significant bioconcentration on ascending food chains, and this is borne out by a summary of their levels in the ambient environment (Table 2).

3 Environmental Budgets

Background and Limitations

In essence, establishing an environmental budget involves quantifying and ranking different environmental compartments as reservoirs of a given pollutant within a defined section of the environment, such as an individual country. The basic principle of an environmental budget is the derivation of a representative concentration for each environmental compartment considered (*e.g.* $10 \, \mu g \, kg^{-1}$ of soil), and its multiplication by an estimate of the volume occupied by that compartment. Whilst obtaining an accurate estimate of compartment volume is not as easy as it would at first appear (requiring answers to such questions as: to what depth are relatively immobile pollutants like PCDD/Fs and PCBs incorporated in soils, and what is the volume occupied by a compartment as loosely defined as 'biota'?), much of the uncertainty involved in environmental budgets is due to problems in deriving representative pollutant concentrations. To illustrate, several attempts have been made to construct environmental

[1] T. Yakushiji, I. Watanabe, R. Kuwabara, T. Tanaka, N. Kashimoto, N. Kunita and I. Hara, *Arch. Environ. Contam. Toxicol.*, 1984, **13**, 341.

2

Table 1 Physicochemical properties of selected PCDD/Fs and PCBs at ambient temperature (290–298 K)[2-5]

Congener	Vp (Pa)[a]	$\log K_{ow}$[b]	W_s (mol m^{-3})[c]
2,3,7,8-TCDD	6.2×10^{-7}	6.8	1.50×10^{-6}
1,2,3,7,8-PeCDD	5.8×10^{-8}	7.4	3.37×10^{-7d}
1,2,3,4,7,8-HxCDD	5.1×10^{-9}	7.8	1.12×10^{-8}
1,2,3,4,6,7,8-HpCDD	7.5×10^{-10}	8.0	5.64×10^{-9}
OCDD	1.1×10^{-10}	8.2	8.70×10^{-10}
2,3,7,8-TCDF	2.0×10^{-6}	6.1	1.37×10^{-6}
2,3,4,7,8-PeCDF	3.5×10^{-7}	6.5	6.93×10^{-7}
1,2,3,4,7,8-HxCDF	3.2×10^{-8}	7.0	2.2×10^{-8}
1,2,3,4,6,7,8-HpCDF	4.7×10^{-9}	7.4	3.30×10^{-9}
OCDF	5.0×10^{-10}	8.0	2.70×10^{-9}
2,2',5,5'-TCB	4.9×10^{-3}	5.84	1.03×10^{-4}
2,2',4,5,5'-PeCB	1.09×10^{-3}	6.38	3.06×10^{-5}
2,2',4,4',5,5'-HxCB	1.19×10^{-4}	6.92	2.8×10^{-6}
2,2',3,3',4,4',6-HpCB	2.73×10^{-5}	6.7	5.06×10^{-6}

[a]PCDD/F and PCB vapour pressures taken from references 2 and 3, respectively.
[b]All $\log K_{ow}$ values obtained from reference 4, except that for 2,2',4,4',5,5'-HxCB which is from reference 3.
[c]PCDD/F and PCB water solubilities taken from references 5 and 3, respectively.
[d]No value available for 1,2,3,7,8-PeCDD; value cited is for 1,2,3,4,7-PeCDD.

budgets for both PCDD/Fs and PCBs. In each case, the accuracy of such efforts is restricted by the extremely limited database relating to concentrations in different environmental media and spatial variations in such concentrations. With regard to spatial variations, information regarding concentrations of these compounds in rural and remote locations is especially scarce. The significant temporal variations in PCB concentrations reported by some authors[11,17] also

2 B. F. Rordorf, *Chemosphere*, 1989, **18**, 783.
3 W. Y. Shiu and D. Mackay, *J. Phys. Chem. Ref. Data*, 1986, **15**, 911.
4 M. S. McLachlan, *Environ. Sci. Technol.*, 1996, **30**, 252.
5 United States Environmental Protection Agency, *Estimating Exposure to Dioxin-Like Compounds*, vol. II, Draft Report, circulated for comment, EPA/600/6-88/005Cb, US EPA, Washington, 1994.
6 ERM/HMIP, *Risk Assessment of Dioxin Releases from MWI Processes*, HMIP, London, 1996.
7 P. Clayton, B. J. Davis, K. C. Jones and P. Jones, *Toxic Organic Micropollutants in Urban Air*, Warren Spring Laboratory Report No. LR904, Warren Spring Laboratory, Stevenage, UK, 1992.
8 C. Rappe, L.-O. Kjeller and S. E. Kulp, in *Proceedings of Dioxin '90-EPRI SEMINAR*, ed. O. Hutzinger and H. Fielder, Ecoinforma Press, Bayreuth, Germany, 1990, p. 207.
9 J. E. Baker, P. D. Capel and S. J. Eisenreich, *Environ. Sci. Technol.*, 1986, **20**, 1136.
10 L.-O. Kjeller, K. C. Jones, A. E. Johnston and C. Rappe, *Environ. Sci. Technol.*, 1991, **25**, 1619.
11 R. E. Alcock, A. E. Johnston, S. P. McGrath, M. L. Berrow and K. C. Jones, *Environ. Sci. Technol.*, 1993, **27**, 1918.
12 Ministry of Agriculture, Fisheries and Food, *Dioxins in Food*, Food Surveillance Paper No. 31, HMSO, London, 1992.
13 J. Mes, W. H. Newsome and H. B. S. Conacher, *Food Additives Contam.*, 1991, **8**, 351.
14 R. Duarte-Davidson, S. J. Harrad, S. Allen, A. S. Sewart and K. C. Jones, *Arch. Environ. Contam. Toxicol.*, 1993, **24**, 100.
15 S. J. Harrad, R. Duarte-Davidson and K. C. Jones, *Organohalogen Compds.*, 1993, **13**, 19.
16 R. Duarte-Davidson, S. V. Wilson and K. C. Jones, *Environ. Pollut.*, 1993, **84**, 69.
17 K. C. Jones, G. Sanders, S. R. Wild, V. Burnett and A. E. Johnston, *Nature*, 1992, **356**, 137.

Table 2 Environmental levels of selected PCDD/Fs and PCBs[6][16]

Congener	Concentration in				
	Urban air (pg m^{-3})[a]	Freshwater (ng m^{-3})[b]	Soil (µg kg^{-1})[c]	Meat (ng kg^{-1})[d]	Human fat (µg kg^{-1})[e]
2,3,7,8-TCDD	6.1×10^{-3}	3.1×10^{-3}	1.1×10^{-4}	0.28	6.4×10^{-3}
1,2,3,7,8-PeCDD	2×10^{-2}	4×10^{-4}	3.0×10^{-4}	0.41	2.3×10^{-2}
1,2,3,4,7,8-HxCDD	1.8×10^{-2}	6×10^{-4}	3.7×10^{-4}	0.55	3.7×10^{-2}
1,2,3,4,6,7,8-HpCDD	0.31	3.6×10^{-2}	6.3×10^{-3}	7.7	0.15
OCDD	1.0	0.14	2.4×10^{-2}	13	0.82
2,3,7,8-TCDF	3.3×10^{-2}	2.6×10^{-2}	9.5×10^{-4}	0.32	9.0×10^{-3}
2,3,4,7,8-PeCDF	2.2×10^{-2}	8.5×10^{-3}	9.3×10^{-4}	0.39	2.4×10^{-2}
1,2,3,4,7,8-HxCDF	7.3×10^{-3}	9.5×10^{-3}	1.3×10^{-3}	0.37	2.6×10^{-2}
1,2,3,4,6,7,8-HpCDF	0.23	9.9×10^{-2}	4.1×10^{-3}	0.5	3.4×10^{-2}
OCDF	0.24	0.1	4.6×10^{-3}	11	4.6×10^{-2}
2,2',5,5'-TCB	110	69	1.1	46	1.8
2,2',4,5,5'-PeCB	68	64	1.8	29	2.3
2,2',3,4,4',5'-HxCB	23	40	1.0	62	120
2,2',3,4,4',5,5'-HpCB	20	–	0.69	39	260

[a]PCDD/F and PCB air concentrations are means reported in references 6 and 7, respectively.
[b]PCDD/F and PCB freshwater concentrations are taken from references 8 and 9, respectively.
[c]PCDD/F and PCB soil concentrations are taken from references 10 and 11, respectively.
[d]PCDD/F and PCB meat concentrations are fresh weight values reported in references 12 and 13, respectively.
[e]PCDD/F concentrations are from reference 14, except for 2,3,7,8-TCDD and 2,3,7,8-TCDF from reference 15; PCB levels in human fat are from reference 16.

hamper efforts to construct a meaningful budget, and budgets conducted using data recorded over a number of years may be subject to significant inaccuracies.

To illustrate the difficulties in deriving representative mean PCB concentrations in an environment as heterogeneous as the open ocean, whilst Tanabe[18] cited a ΣPCB concentration of 0.6 ng dm^{-3} in North Atlantic seawater, Harrad et al.[19] employed a value for North Atlantic and North Sea seawater (levels in this latter area significantly exceeded the former) of 0.12 ng dm^{-3}. Although Harrad et al. noted the higher estimates of others,[20] they suggested that their own concentration estimate may have been too high, taking into account the concentration decline observed with increased sampling depth,[21] and the fact that the bulk of the samples on which their estimate was based were taken only 6 m below the surface.

Clearly, the derivation of representative concentrations for each of the environmental compartments considered is crucial to the accuracy of any environmental budget. To illustrate, whilst Harrad et al.[19] calculated the ΣPCB burden of a seawater volume of 1.14×10^{17} dm^3 (including the North Sea) to be 14 t, Lohse[20] used a representative concentration of 3.5 ng ΣPCB dm^{-3} to derive a seawater ΣPCB loading of 150 t for the North Sea alone, a volume of 5.25×10^{16} dm^3. Such significantly different conclusions concerning the burden of a comparatively well-characterized location illustrates the extent of uncertainty associated with the construction of budgets, and particularly their dependence on accurate concentration data.

Despite these limitations, the construction of environmental budgets plays an important rôle in efforts to understand the environmental fate and behaviour of PCDD/Fs and PCBs, and the following section will examine a selection of the most detailed conducted to date.

PCDD/Fs

Harrad and Jones[22] constructed a budget for the terrestrial UK environment. Unfortunately, although freshwater and freshwater sediments were considered, the absence of sufficiently detailed data relating to PCDD/F contamination of terrestrial biota and the marine environment, as well as the difficulty in deriving a representative concentration for an environmental compartment as diverse in composition as biota, meant that the significance of these potentially important reservoirs of PCDD/Fs was not quantified. The findings of this exercise (summarized in Table 3) were that, within the UK, topsoil represents easily the most important reservoir for tetra- to octachlorinated dioxins and furans, with other compartments such as freshwater sediments, ambient air, freshwater and vegetation making comparatively negligible contributions to the overall burden.

[18] S. Tanabe, *Environ. Pollut.*, 1988, **50**, 5.
[19] S. J. Harrad, A. P. Sewart, R. Alcock, R. Boumphrey, V. Burnett, R. Duarte-Davidson, C. Halsall, G. Sanders, K. Waterhouse, S. R. Wild and K. C. Jones, *Environ. Pollut.*, 1994, **85**, 131.
[20] J. Lohse, PhD Thesis, Universität Hamburg, 1988.
[21] D. E. Schulz, G. Petrick and J. C. Duinker, *Mar. Pollut. Bull.*, 1988, **19**, 526.
[22] S. J. Harrad and K. C. Jones, *Sci. Total Environ.*, 1992, **126**, 89.

Table 3 Summary of UK
PCDD/F budget of
Harrad and Jones[22]

Compartment	Loading (kg)[a]	% of total loading
Topsoil (top 5 cm)	5560	99.5
Freshwater sediments (top 5 cm)	25.2	0.45
Vegetation	2.9	0.05
Humans	*ca*.1	<0.02
Air	0.85	<0.02
Freshwater	0.088	<0.02

[a]Loadings expressed as the sum of all tetra- to octachlorinated PCDD/Fs.

PCBs

The comparative ease of PCB measurement has generated a relatively detailed database relating to the presence of these compounds in the environment. As a result, the distribution of ΣPCB and PCBs number 28, 52, 101, 138, 153 and 180 has been considered within an area encompassing the UK surface and the marine environment within a 200 km perimeter zone around the UK shoreline.[19] The findings of this study (summarized in Table 4) are that topsoil contains the majority of the ΣPCB burden within the area considered, with other significant reservoirs identified as seawater and marine sediments.

For comparison, Table 5 lists the findings of Tanabe,[18] who calculated the ΣPCB burden in the global environment. He estimated the global burden to total 374 000 t of ΣPCB (31% of total world PCB production), with the overwhelming majority (96%) associated with seawater and coastal sediments. This estimate compares favourably with that of the United States National Academy of Sciences, who calculated that the oceanic water over the North American basin held 66 000 t ΣPCB.[23] Overall the work of Tanabe and coworkers[18,24] indicates the open ocean to contain around 61% and the terrestrial and coastal environment 39% of the global environmental burden of PCBs.

There are two apparent discrepancies between the two studies mentioned above. First is the fact that Harrad *et al.*[19] identified topsoil as the major environmental reservoir of PCBs, whilst Tanabe[18] pinpointed seawater as bearing the most significant fraction of the environmental burden. This difference is mainly attributable to the fact that although the land:ocean surface area ratio of the UK environment considered by Harrad *et al.*[19] is only slightly higher (at 0.44) than that which prevails for the Earth as a whole (*i.e.* 0.41, assuming global land and ocean surface areas of 1.48×10^{14} and 3.62×10^{14} m^2, respectively[25]), the ocean depth (200 m) assumed by Harrad *et al.* is considerably less than that assumed by Tanabe (3729 m). As a result, the land:ocean volume ratio of the area considered by Harrad *et al.* is significantly below that of the Earth overall.

The second apparent discrepancy is that whilst Harrad *et al.* calculated that the

[23] Environmental Studies Board, in *Polychlorinated Biphenyls*, National Academy of Sciences, Washington, 1979.

[24] S. Tanabe and R. Tatsukawa, in *PCBs and the Environment*, Vol. 1, ed. J. S. Waid, CRC Press, Florida, 1986, p. 143.

[25] K. Ballschmiter, *Angew. Chem., Int. Ed. Engl.*, 1992, **31**, 487.

Table 4 Summary of UK PCB budget of Harrad et al.[19]

Compartment	Loading (t) of congener number in (% of total loading)							
	28	52	101	138	153	180	ΣPCB
Topsoil	21 (87.1)	14 (88.7)	22 (91)	12 (81.6)	19 (90.6)	8.6 (82.3)	370 (93.1)
Seawater	0.67 (2.8)	0.66 (4.2)	1.3 (5.4)	1.6 (10.9)	0.78 (3.7)	0.86 (8.2)	14 (3.5)
Marine sediments	1.6 (6.6)	0.82 (5.2)	0.74 (3.1)	0.89 (6.1)	1.0 (4.8)	0.6 (5.7)	8.2 (2.1)
Freshwater sediments	0.18 (0.8)	0.09 (0.5)	0.11 (0.5)	0.09 (0.6)	0.07 (0.3)	0.16 (1.5)	2.2 (0.6)
Vegetation	0.57 (2.4)	0.18 (1.1)	–	<6.5 (<0.1)	<9.8 (<0.1)	–	1.8 (0.5)
Humans	8×10^{-3} (<0.1)	2×10^{-3} (<0.1)	2×10^{-3} (<0.1)	0.10 (0.7)	0.11 (0.5)	0.21 (2.0)	0.76 (0.2)
Sewage sludge	0.02 (0.1)	0.02 (0.2)	0.02 (0.1)	0.01 (0.1)	0.02 (0.1)	0.02 (0.2)	0.48 (0.1)
Air	5×10^{-3} (<0.1)	5×10^{-3} (<0.1)	5×10^{-3} (<0.1)	3×10^{-3} (<0.1)	3×10^{-3} (<0.1)	1×10^{-3} (<0.1)	0.065 (<0.1)
Freshwater	1×10^{-3} (<0.1)	2×10^{-3} (<0.1)	2×10^{-3} (<0.1)	1×10^{-3} (<0.1)	6×10^{-4} (<0.1)	–	0.03 (<0.1)
Total loading	24	16	24	15	21	11	400

	Compartment	Loading (t)[a]	% of total loading
Table 5 Summary of global PCB budget of Tanabe[18]	*Terrestrial & Coastal*		
	Air	500	0.13
	Freshwater	3500	0.94
	Sediment	130 000	35
	Soil	2400	0.64
	Seawater	2400	0.64
	Biota	4300	1.1
	Oceanic		
	Air	790	0.21
	Seawater	230 000	61
	Sediment	110	0.03
	Biota	270	0.07
	Total	374 000	100

[a]Loadings expressed as ΣPCB.

overall ΣPCB burden for the area considered amounted to 400 t (*ca.* 1% of total UK sales of PCBs), Tanabe estimated that 31% of world PCB production had been released into the environment. More reasonable agreement between the two studies emerges when UK archived soil concentration data[11] are used to calculate the burden of ΣPCB in UK topsoil at the peak of PCB contamination. The maximum total burden was estimated to be 23 200 t (over 50% of total UK sales), a figure that correlates well with that of Tanabe. However, whilst it would appear probable that a significant fraction of PCB production has already 'escaped', it is equally apparent that there remains a significant fraction of PCBs with the potential for future release into the environment. Tanabe concluded that the bulk (65%) of world PCB production remained in use, or was 'locked' in landfills and hazardous waste dumps, and thus identified an urgent need to develop technologies capable of destroying such land-locked PCBs before they were released into the environment.

Comparison of the peak PCB burden with the present loading for the UK suggests that a significant fraction of UK PCB sales were released into the environment, but have since 'disappeared' from the UK. Firm evidence of the fate of these 'lost' PCBs is not available, but the most likely loss mechanisms are a combination of anthropogenic and natural degradation, along with atmospheric and pelagic transport away from the UK. Any assessment of the relative importance of these processes must remain largely speculative in the absence of unequivocal supporting data, but comparisons of the relative PCB losses likely from soils due to biodegradation and volatilization indicate that only the latter is likely to be able to account for the dramatic decline in the UK PCB burden over the last two decades. This conclusion is supported by reports that ^{210}Pb corrected PCB fluxes to the Agassiz ice cap in Ellesmere Island, Canada, remained essentially constant over the period 1976/77 to 1985/86, at a level approximately 50% of that observed between 1970/71 and 1975/76.[26] By

[26] D. J. Gregor, in *Seasonal Snowpacks*, NATO ASI Series, vol. G 28, ed. T. D. Davies, Springer, Berlin, 1991, p. 323.

comparison, levels in soils from several UK sites are reported to have fallen by a significantly greater factor (in 1984, the ΣPCB soil concentration at Woburn was *ca.* 10% of that in 1972) over a similar time period.[11] Similar rates of decline have been observed in other UK abiotic environmental compartments like grass and air.[17,27] Such observations lend credence to the hypothesis that PCB volatilization rather than degradation is the principal loss mechanism from the UK, otherwise the rate of decline would have been roughly the same in all locations. This concurs with the finding that by the early 1980s, just 3.6% of US PCB production had been degraded or incinerated.[28]

When coupled with the detection of elevated PCB levels in biota from remote polar locations,[29–31] the suggestions of large-scale volatilization of PCBs from temperate industrialized nations have spawned the 'Global Distillation' hypothesis.[32] The central tenet of this hypothesis is that semi-volatile organic compounds (SVOCs), like PCBs, volatilize from warm and temperate locations, and subsequently undergo long-range atmospheric transport throughout the globe. Following deposition in polar regions, the extreme low temperatures in such regions minimize subsequent volatilization, with the result that over time one would expect a shift of the global PCB loading from temperate industrialized nations to both the Arctic and Antarctic.

4   Source Inventories

Background and Limitations

The production of source inventories is essentially the process of identifying and ranking sources of a given pollutant to a given section of the environment. Such rankings permit the identification of major release pathways, and hence the prioritization of emission control policies. The basic strategy of a source inventory is to derive an emission factor for a specific source activity (*e.g.* 10 μg per t of waste burnt), and subsequently to multiply this by an activity factor, *i.e.* the extent to which the activity is practised (*e.g.* 3 million t waste burnt per year). Generally, activity factors are far more reliable than emission factors (the latter can vary considerably) and uncertainties surrounding the latter are the principal source of inaccuracies in source inventories. In recognition of such problems, it is now becoming commonplace to assess the 'quality' of both the emission and activity factors used. Such 'quality' assessments (based, for example, on the number and range of concentration values used to derive an emission factor) are unavoidably subjective, but are still to be considered as a welcome development in what remains an extremely important but somewhat 'inexact' area of research.

In light of the uncertainties associated with source inventories, it is important that their validity is evaluated as far as possible. One way in which this may be

27   K. C. Jones, R. Duarte-Davidson and P. A. Cawse, *Environ. Sci. Technol.*, 1995, **29**, 272.

28   S. Miller, *Environ. Sci. Technol.*, 1982, **16**, 98A.

29   R. J. Norstrom, M. Simon, D. C. G. Muir and R. E. Schweinsberg, *Environ. Sci. Technol.*, 1988, **22**, 1063.

30   E. Dewailly, A. Nantel, J.-P. Weber and F. Meyer, *Bull. Environ. Contam. Toxicol.*, 1989, **43**, 641.

31   D. C. G. Muir, M. D. Segstro, K. A. Hobson, C. A. Ford, R. E. A. Stewart and S. Olpinski, *Environ. Pollut.*, 1995, **90**, 335.

32   F. Wania and D. Mackay, *Ambio*, 1993, **22**, 10.

achieved is the comparison of estimated total annual atmospheric emissions with estimated annual atmospheric deposition. For compounds possessing the environmental stability of PCDD/Fs and PCBs, one would in theory expect the two estimates to concur, with any shortfall in atmospheric emissions deemed indicative of a flaw in the source inventory. Of course, even if both emissions and deposition were perfectly characterized, one would not expect total agreement, as both PCDD/Fs and PCBs are subject to environmental degradation. Furthermore, national or regional source inventories fail to consider the influence of environmental transport of emitted material away from the nation/region of concern, as well as the potential 'import' of material emitted outside the boundaries considered by the source inventory. Indeed, the latter is the most probable explanation for the presence of PCBs in UK soils prior to the onset of large-scale UK use of these compounds in 1954. Aside of these potential sources of error, there exist possible problems with the derivation of accurate emission factors (discussed above), as well as difficulties in procuring a representative estimate of depositional inputs over a large area, when, for example, what little data are available indicate PCDD/F deposition to be far greater in urban than rural locations.[6] Furthermore, it is by no means certain that current methods of sampling atmospheric deposition ensure 100% collection efficiency, with resultant uncertainty in estimates of depositional input. However, in spite of such problems, it is considered that the comparison of emission and deposition estimates is a worthwhile exercise, and its applications to both PCBs and PCDD/Fs are discussed in the following sections.

PCBs

Little doubt exists as to the principal source of the present environmental burden of PCBs, of which an estimated 1.2 million t were manufactured worldwide between the onset of their production in 1929 and the late 1970s, when their production in most western nations ceased.[33] Despite this, there remains interest in identifying the sources of the continuing input of these compounds. Harrad *et al.*[19] noted the conclusion of Eduljee[34] that, in the absence of fresh production, volatilization of previously deposited material constituted the principal source of continuing PCB inputs, and using a theoretical treatment of PCB volatilization[35] estimated the extent of such volatilization from topsoil as part of an attempt to construct a PCB source inventory for the UK. Their findings, summarized in Table 6, indicate revolatilization of previously deposited material from topsoil to be the largest current source of PCBs to the UK atmosphere, with other significant sources including leaks from PCB-filled transformers and capacitors remaining in service.

Emissions versus Deposition. Current PCB emissions to the UK atmosphere have been estimated to be around twice the current depositional inputs.[19] Whilst

[33] J. D. Bletchly, in *Proceedings of the PCB Seminar*, ed. M. C. Barros, H. Könemann and R. Visser, Ministry of Housing, Physical Planning and Environment, The Hague, The Netherlands, 1983, p. 343.

[34] G. H. Eduljee, *Chem. Br.*, 1988, **24**, 241.

[35] G. H. Eduljee, *Chemosphere*, 1987, **16**, 907.

Table 6 Summary of UK PCB source inventory of Harrad et al.[19]

Compartment	Estimated atmospheric releases (kg yr^{-1}) of congener number in (% of total releases)						
	28	52	101	138	153	180	ΣPCB
Volatilization from soil	5100 (93.7)	4000 (95.0)	2400 (97.4)	840 (96.2)	1000 (96.5)	680 (98.4)	40000 (89.9)
Capacitor leaks[a]	300 (5.5)	180 (4.3)	41 (1.7)	20 (2.3)	19 (1.8)	1.7 (0.3)	3900 (8.8)
Transformer leaks[a]	6 (0.1)	8 (0.2)	13 (0.5)	8.5 (1.0)	14 (1.4)	6.5 (0.9)	250 (0.6)
Scrap metal recovery	26 (0.5)	11 (0.2)	7.6 (0.3)	3.2 (0.4)	2.0 (0.2)	0.6 (0.1)	240 (0.5)
Volatilization from land-applied sewage	5.7 (0.1)	12 (0.3)	3.5 (0.1)	1.5 (0.1)	1.3 (0.1)	2.1 (0.3)	85 (0.2)
Total	5400	4200	2500	870	1000	690	45000

[a]Values quoted are the mean of the range cited in original paper.

the database on which this conclusion was made was limited, *e.g.* estimates of PCB volatilization were based on theoretical calculations only, and depositional input was derived from only 6 months data for two urban locations, the inference is that there is presently a net flux of PCBs out of the UK. This conclusion is supported by other workers,[36] who have reported that water bodies such as Lake Superior now constitute a net source of PCBs to the atmosphere as a consequence of outgassing, and lends further support to the 'Global Distillation' hypothesis.

One factor overlooked in the comparison of UK PCB emissions and deposition[19] was the fact that whilst calculations of ΣPCB emissions were genuinely based on ΣPCB, depositional input of 'ΣPCB' was based on only a limited number (44) of PCB congeners. This is significant, as other researchers have reported the number of PCB congeners present in the environment to be around 100, and the 44 congeners on which the depositional input data of Harrad *et al.* were based[19] exclude congeners like PCB 31, which are usually present in equal concentration to PCB 28 in abiotic matrices.[21,37] The omission of such congeners will inevitably lead to an underestimation of the depositional flux of ΣPCB, thus exaggerating the discrepancy between emissions and deposition, and it would appear prudent to monitor all environmentally present PCBs in future studies.

PCDD/Fs

Unlike PCBs, PCDD/Fs have never been deliberately manufactured, other than on a laboratory scale for use as analytical standards. Instead, it is generally accepted that their omnipresence has arisen from their formation during a variety of anthropogenic combustion activities and industrial processes such as the manufacture and use of organochlorine chemicals.[38] It is generally agreed that, whilst some natural sources of PCDD/Fs exist, like the enzymatic self-condensation of chlorophenols,[39] and forest fires,[40] the contribution of such sources to the contemporary environmental PCDD/F loading is minimal. This conclusion is supported by the analysis of a variety of archived environmental materials, which reveal significantly higher PCDD/F contamination in the latter half of this century than in the past.[10] Despite these areas of agreement, the relative significance of the many potential sources of PCDD/Fs remains uncertain, and as a result the origins of the ubiquitous environmental presence of these compounds have been the subject of considerable study, and the number of PCDD/F source inventories far exceeds those for PCBs. Table 7 summarizes the findings of a selection of the PCDD/F source inventories conducted to date.

One common factor of these and other PCDD/F source inventories is that they are limited to the consideration of primary sources of atmospheric emissions.

[36] K. C. Hornbuckle, J. D. Jeremiason, C. W. Sweet and S. J. Eisenreich, *Environ. Sci. Technol.*, 1994, **28**, 1491.

[37] R. Wittlinger and K. Ballschmiter, *Chemosphere*, 1987, **16**, 2497.

[38] S. J. Harrad and K. C. Jones, *Chem. Br.*, 1992, **28**, 1110.

[39] A. Svenson, L.-O. Kjeller and C. Rappe, *Environ. Sci. Technol.*, 1989, **23**, 900.

[40] R. E. Clement and C. Tashiro, *Forest Fires as a Source of PCDD and PCDF*, presented at the 11th International Symposium on Chlorinated Dioxins and Related Compounds, Research Triangle Park, NC, USA, 1991.

Table 7 Summary of selected national PCDD/F source inventories[22,41,42]

Source	Annual atmospheric release (kg yr^{-1}) estimated by		
	Hutzinger & Fiedler[41] [a]	Harrad & Jones[22] [b]	Eduljee & Dyke[42] [c]
Municipal waste incineration	0.22	10.9	0.52
Chemical waste incineration	0.039	'minimal'	5.1×10^{-3}
Clinical waste incineration	0.054	1.7	0.053
Industrial coal combustion	–	7.7	0.036
Domestic coal combustion	4.1×10^{-3} [d]	5.1	0.027
Leaded petrol combustion	7.2×10^{-3}	0.7	–
Unleaded petrol combustion	8.0×10^{-4}	–	–
Diesel fuel combustion	4.6×10^{-3}	–	–
Traffic	–	–	0.023
Chlorophenol usage	–	1.7[e]	8×10^{-4}

[a] Release expressed in terms of Σi-TE for Germany.
[b] Release expressed in terms of Σtetra- to octachlorinated PCDD/Fs for the UK.
[c] Release expressed in terms of Σi-TE for the UK; where necessary, values quoted are the mean of the range cited in original paper.
[d] includes domestic combustion of oil, pit-coal, coke and 'brickets'.
[e] Volatilization from chlorophenol-treated substrates.

Unfortunately, this overlooks the potentially significant contribution of secondary sources of atmospheric emissions, such as the remobilization of the existing topsoil burden, and omits consideration of other non-atmospheric pathways *via* which PCDD/Fs may be released into the environment. Given that discharges to the aquatic environment are disproportionately significant for non-combustion sources like the manufacture and use of organochlorine biocides such as chlorophenols used for timber treatment and the chlorine bleaching of raw paper pulp, most source inventories fail to consider fully the impact of such non-combustion sources. As an illustration of the potential contribution made by non-combustion sources to the environmental burden of PCDD/Fs, it has been estimated that release of just 1% of the Σ tetra- to octa-PCDD/Fs associated with the annual quantity of tetra- and pentachlorophenols used in the UK would amount to 17 kg yr^{-1}, more significant than any of the individual combustion source categories considered.[22] Similarly, a source inventory for the Austrian environment estimated the cumulative release of PCDD/Fs as a result of the manufacture and use of organochlorine chemicals to amount to 12 kg Σi-TE over 30 years, which was compared to a contemporary annual emission figure of 80 g Σi-TE from other sources (primarily combustion activities).[43] Overall, therefore, it would appear that combustion and non-combustion activities rank roughly equal as sources of the present environmental burden of ΣPCDD/Fs. However, as discussed elsewhere,[22] combustion activities such as waste incineration and fossil fuel combustion probably constitute a far greater source of tetra- and penta-PCDD/Fs than the manufacture and use of chlorophenols, which in turn represent the greater source of hepta- and octa-PCDD/Fs.

[41] O. Hutzinger and H. Fielder, *Chemosphere*, 1993, **27**, 121.
[42] G. H. Eduljee and P. Dyke, *Sci. Total Environ.*, 1996, **177**, 303.
[43] A. Riss and H. Aichinger, *Organohalogen Compds.*, 1993, **14**, 341.

Table 8 Comparisons of PCDD/F atmospheric emissions with deposition[5,22,42,45,46]

Country	Emission:deposition ratio
United States[45]	0.11[a]
United States[5]	0.07–1.3[b]
United Kingdom[22]	0.12[c]
United States[46]	<0.48[b]
United Kingdom[42]	1.1–2.2[b]

[a]Calculated in terms of 2,3,7,8-TCDD.
[b]Calculated in terms of Σi-TE.
[c]Calculated in terms of Σtetra- to octa-PCDD/Fs.

Emissions versus Deposition. Several workers have compared estimates of atmospheric emissions with measured deposition fluxes for PCDD/Fs; a selection are summarized in Table 8. Whilst the database on which such comparisons are founded is limited, the tentative conclusions which may be derived are of potentially great significance. In particular, it has been observed that most such comparisons, including several of those referred to in Table 8, indicate that current deposition exceeds atmospheric emissions from primary sources.[44] Whilst there are significant sources of potential error involved in such calculations (*e.g.* in addition to those mentioned in earlier sections, like the failure to consider releases from secondary sources, there is the possibility that estimates of representative depositional inputs may overestimate the true figure, owing to the fact that measurements of rural depositional inputs are extremely scarce), the obvious conclusion is that we have yet to discover all sources of the current depositional inputs of PCDD/Fs.

In contrast to earlier inventories, that of Eduljee and Dyke[42] obtains a far closer correlation between emissions and deposition. Certainly, theirs is probably the most comprehensive PCDD/F source inventory yet compiled, and its quantification of many previously known but uncharacterized sources explains much of the discrepancy reported by previous inventories. However, their estimate of the overall depositional flux to the UK appears extremely low, and was based on a very limited number of rural measurements. Furthermore, their emissions inventory and comparisons with depositional input are based on PCDD/Fs expressed as Σi-TE. The use of such a unit presents a potential problem when comparing emissions and deposition, as environmental weathering processes transform PCDD/F source profiles into an environmental congener pattern favouring the higher chlorinated congeners, which make considerably lower contributions to Σi-TE values. As a result, comparisons of PCDD/F emissions with deposition conducted on a Σi-TE basis will tend to provide a higher emission:deposition ratio than would be obtained were the same comparison conducted on a congener-specific or a ΣPCDD/F (either total homologue, or total 2,3,7,8-chlorinated congener) basis. Such non-Σi-TE approaches were utilized by two of the studies included in Table 8,[22,45] and these differences

[44] C. Rappe and L.-O. Kjeller, *Organohalogen Compds.*, 1994, **20**, 1.
[45] C. C. Travis and H. A. Hattemer-Frey, *Sci. Total Environ.*, 1991, **104**, 97.
[46] J. Schaum, D. Cleverly, M. Lorber, L. Phillips and G. Schweer, *Organohalogen Compds.*, 1993, **14**, 319.

in the units used to report PCDD/F emissions and deposition may well explain at least some of the apparent conflict between different researchers.

As a result of this uncertainty, it is thus imperative to establish whether there is any discrepancy between atmospheric emissions and deposition of PCDD/Fs, and, if so, to quantify its extent. If this goal is to be achieved, then the frequency and spatial variation of both deposition measurements, particularly in remote areas, and emission factors for both established and suspected PCDD/F sources, including the significance of remobilization of the existing burden, must be increased.

5 Conclusions

Budget and source inventories provide useful indicators of areas requiring further research. This review has identified two particularly important topics where our understanding of the sources and environmental fate of PCDD/Fs and PCBs is limited. First, based on a detailed budget for PCBs in the global ocean environment, and interpretation of archived soil data for PCBs, it may be concluded that a significant fraction of PCB production has already been released into the environment. If this true, then it is imperative that the fate of these 'lost' PCBs is ascertained. In particular, the relative significance of the rôle played by environmental degradation, along with the magnitude of both the open oceans and polar regions as PCB reservoirs, must be determined. Second, the apparent discrepancy between different PCDD/F source inventories requires clarification. Specifically, whilst the atmospheric emissions quantified by some can only account for *ca.* 10% of current atmospheric inputs, others appear to be able to account for all atmospheric deposition. This is an important issue, as the former situation suggests that urgent efforts are required to identify the major sources of PCDD/Fs to the atmosphere, in order that future inputs may be reduced.

If these uncertainties are to be resolved, it is clear that the databases on which both budgets and source inventories are based must be improved. For budgets, there exists considerable uncertainty over the concentrations used to estimate loadings in different environmental compartments. The only way in which such uncertainty may be reduced is by increasing the extent and scope of environmental monitoring, in order to enhance our knowledge of both inter- and intra-compartmental concentration variability. It is particularly important to ascertain intra-compartmental spatial variations, *e.g.* the relationship between seawater contamination and both sampling depth and geographical location. With regard to source inventories, the uncertainty associated with emission factors remains considerable, and it is considered vital that the number of measurements on which emission factors are founded is increased.

Human Exposure to Dioxin

CURTIS C. TRAVIS AND APRIL G. NIXON

1 Introduction

Dioxin is the generic name for a family of highly toxic chlorinated compounds produced as unwanted by-products of combustion and manufacturing processes. The high toxicity of dioxin has led the scientific community, the media and the public to focus increasing attention on the nature and extent of human exposure to dioxin. Although there are 75 congeners (chemical compounds) of polychlorinated dibenzo-*p*-dioxins (PCDDs) and 135 congeners of polychlorinated dibenzofurans (PCDFs), the term 'dioxin' is generally used to refer to 2,3,7,8-tetrachlorodibenzo-*p*-dioxin (TCDD), the most potent chemical toxin ever evaluated by the United States Environmental Protection Agency (EPA). Although dioxin may be produced naturally (in forest fires, for example), the greatest concern is for anthropogenic sources, such as municipal solid waste incinerators, medical waste incinerators, pulp and paper mills, motor vehicles and wood burning (both residential and industrial).

The goal of this chapter is to identify the major sources of dioxin and related compounds, to identify the major pathways of human exposure and to estimate the extent of the resulting human exposure.

2 Measuring Toxicity Levels

Since dioxins and furans have varying levels of toxicity, emissions of mixtures of these compounds are typically expressed in TEQs (Toxic EQuivalents). TEQs relate the toxicity of all dioxin and furan compounds to the known toxicity of 2,3,7,8-TCDD using a weighting scheme adopted by the EPA and most European countries.[1,2] Therefore, a quantity of combined PCDDs and PCDFs

[1] US Environmental Protection Agency, *Interim Procedures for Estimating Risks Associated with Exposures to Mixtures of Chlorinated Dibenzo-*p*-dioxins and Dibenzofurans (CDDs and CDFs) and 1989 Update*, 22161 PB90-145756, US Department of Commerce, National Technical Information Service, Springfield, VA, 1989.

[2] NATO, *Pilot Study on International Information Exchange on Dioxins and Related Compounds: Emissions of Dioxins and Related Compounds from Combustion and Incineration Sources*, North Atlantic Treaty Organization, Committee on the Challenges of Modern Society, Report 172, August 1988.

expressed in TEQs is assumed to have the same toxic effect as an equal quantity of 2,3,7,8-TCDD.

3 Environmental Concentrations Reveal Ubiquity of Dioxin

PCDDs and PCDFs, including the toxic 2,3,7,8-TCDD, have been measured in practically all media, including air, soil, sediment, meat, milk and dairy products, fish and shellfish, vegetation and human biological samples (Table 1).[3-9] These findings clearly indicate that PCDDs and PCDFs are widespread and that virtually everyone – regardless of age, gender or geographic residence – is potentially exposed to these compounds on a daily basis.

4 Sources of Dioxin

A number of combustion and chemical production processes contribute to environmental concentrations of PCDD/F. Sources that have traditionally caused the greatest concern include municipal waste incinerators, hospital waste incinerators, bleached chemical wood pulp and paper mills, motor vehicles and wood combustion. We have attempted to represent the most recent data available on PCDD/F emissions from these sources. It should be remembered that the list presented here is by no means exhaustive. Potential sources of TCDD not discussed in the following paragraphs include discharges from metal processing and treatment plants, copper smelting plants and pentachlorophenol production.

Municipal Solid Waste Incinerators

Municipal solid waste incinerators (MSWIs) are often thought to be the largest source of dioxin in the US.[10] Approximately 176 MSWIs in the US[11] combust 2.9×10^{10} kg yr^{-1} of municipal solid waste.[12] Applying an emission factor of 3.56×10^{-8} g TEQ kg^{-1} of municipal waste[13] yields an estimate of 1032 g

[3] C. Halsall and K. C. Jones, presented at Dioxin '93, 13th International Symposium on Chlorinated Dioxins and Related Compounds, Vienna, Austria, 1993.

[4] US Environmental Protection Agency, *National Dioxin Study*, EPA/530-SW-87-025, Office of Solid Waste and Emergency Response, Washington, 1987.

[5] R. F. Bopp, M. L. Gross, H. Tong, M. J. Simpson, S. J. Monson, B. L. Deck and F. C. Moser, *Environ. Sci. Technol.*, 1991, **25**, 951.

[6] A. Schecter, J. Startin, C. Wright, M. Kelly, O. Päpke, A. Lis, M. Ball and J. Olson, *Chemosphere*, 1994, **29**, 2261.

[7] US Environmental Protection Agency, *National Study of Chemical Residues in Fish*, EPA/832-R-02-008, Office of Science and Technology, Washington, 1992.

[8] A. Hülster and H. Marschner, *Chemosphere*, 1993, **27**, 439.

[9] D. G. Patterson, G. D. Todd, W. E. Turner, V. Maggio, L. R. Alexander and L. L. Needham, *Environ. Health Perspect.*, 1994, **101**, 195.

[10] V. M. Thomas and T. G. Spiro, *Toxicol. Environ. Chem.*, 1995, **50**, 1.

[11] B. K. Burton and J. V. L. Kiser, in *Municipal Waste Combustion, Proceedings of an International Speciality Conference of the Air and Waste Management Association, Williamsburg, VA.*, 1993, p. 3.

[12] American Society of Mechanical Engineers, *The Relationship between Chlorine in Waste Streams and Dioxin Emissions from Waste Combustor Stacks*, CRTD vol. 36, AMSE, New York, 1995.

[13] H. G. Rigo, *Solid Waste Technol.*, 1995, 36.

Table 1 Summary of PCDD/F levels in environmental media and food (whole weight basis)

Media	North America[a]	Europe[a]
Air (pg m^{-3})	$n = 84$	$n = 454$
TEQ	0.095 ± 0.24	0.108
Soil (ppt)	$n = 95$	$n = 133$
TEQ	8.0 ± 5.7	8.7
Sediment (ppt)	$n = 7$	$n = 20$
TEQ	3.9	34.8
Fish (ppt)	$n = 60$	$n = 18$
TEQ	1.16 ± 1.2	0.93
Water (ppq)	$n = 214$	
TEQ	0.0056 ± 0.0079	NDA
Milk	$n = 2$	$n = 168$
TEQ	0.07	0.05
Dairy (ppt)	$n = 5$	$n = 10$
TEQ	0.36 ± 0.29	0.08
Eggs (ppt)	$n = 8$	$n = 1$
TEQ	0.14 ± 0.12	0.15
Beef (ppt)	$n = 14$	$n = 7$
TEQ	0.48 ± 0.99	0.32;[b] 0.61[c]
Pork (ppt)	$n = 12$	$n = 3$
TEQ	0.26 ± 0.13	<0.06
Chicken (ppt)	$n = 9$	$n = 2$
TEQ	0.19 ± 0.29	0.21

Source: EPA (1994 draft).
[a]Arithmetic mean TEQs and standard deviations.
[b]Beef.
[c]Veal.

TEQ yr^{-1} from municipal waste incinerators. It is anticipated that compliance with EPA emissions guidelines issued in 1991 will reduce total dioxin flux from MSWIs to about 230 g yr^{-1}, and that 1994 guidelines could further reduce emissions to approximately 60 g TEQ yr^{-1}.[13]

EPA[14] assumes an emission factor of 1.0×10^{-7} g TEQ kg^{-1} for MSWIs, which yields an estimate of 3000 g TEQ yr^{-1} emitted from MWCs. It is widely believed that the EPA estimate is too high and that around 1000 g TEQ yr^{-1} is a more reasonable estimate.

Hospital Waste Incinerators

Incineration of medical waste at hospitals and other health-care facilities is often cited as a prominent source of PCDD/Fs.[14] A total of 2223 operating hospital incinerators combust 2.6×10^9 kg of medical waste per year.[15] Using an

[14] US Environmental Protection Agency, *Estimating Exposure to Dioxin-like Compounds, Vol. II: Properties, Sources, Occurrence and Background Exposures*, review draft, unpublished, June 1994.
[15] American Hospital Association, *Comments from the American Hospital Association on the United States Environmental Protection Agency Dioxin Exposure and Health Effects Documents*, AHA, Chicago, IL, January 13, 1995.

emission factor of 5.3×10^{-8} g TEQ kg^{-1}, the American Hospital Association (AHA)[15] estimates total emissions from US hospital incinerators at 137.5 g TEQ yr^{-1}.

Based on a review of data from 20 units, ranging from small, uncontrolled units to larger incinerators with sophisticated air pollution control systems, Rigo[13] estimates an emission factor of 1.1×10^{-7} g TEQ kg^{-1}, assuming that US hospital incinerators combust approximately 3.0×10^9 kg of medical waste per year. Rigo estimates total emissions from US hospital incinerators to be 325 g TEQ yr^{-1}.

Although the EPA[14] estimated PCDD/F emissions from medical waste incinerators in 1990 at 5100 g TEQ yr^{-1}, the agency is re-examining its calculations and attempting to resolve discrepancies between its estimate and that submitted by the AHA. Although no new estimate has been officially released, the EPA is expected to reach lower estimated emission rates for medical waste incinerators.

The emission factor for hospital waste incinerators (5.3×10^{-8} g TEQ kg^{-1})[15] is approximately twice as large as those for municipal waste incinerators (3.56×10^{-8} g TEQ kg^{-1};[13] 1.0×10^{-7} TEQ kg^{-1} [14]). This may be due to the fact that (1) hospital waste contains a large amount of plastic and therefore has a higher chlorine concentration than municipal solid waste and (2) hospital incinerators are smaller and therefore may be less well operated or use poorer combustion practices.[10] However, higher chlorine content does not necessarily imply a higher level of PCDD/F emission. A study of chlorine in waste streams and dioxin emissions from waste combustor stacks found no statistically significant relationship between chlorine input and PCDD/F stack gas concentrations for the majority (80%) of the 90 facilities examined.[12] In fact, 9% of the facilities studied displayed decreasing PCDD/F concentrations with increasing chlorine.

Hazardous Waste Incinerators

A total of 190 hazardous waste incinerators (HWIs) exist in the US, combusting approximately 1.3 million metric tons of hazardous waste each year.[14] Applying an emission factor of 2.7×10^{-8} g TEQ kg^{-1} waste feed[16] to the annual waste combustion yields an estimate of 35 g TEQ yr^{-1} emitted from US hazardous waste incinerators (not including cement kilns).[14]

Cement Kilns

Cement kilns, which are fuelled by the combustion of hazardous waste or fossil fuels, are a source of environmental dioxins. A total of 212 cement kilns combust approximately 104 billion kg of raw materials each year; approximately 34 of these kilns burn hazardous waste as supplemental fuel.[17] Emission factors of

---

[16] J. J. Helble, *Analysis of Dioxin Emissions from the Incineration of Hazardous Waste*, US Environmental Protection Agency, Office of Solid Waste, Permits and State Programs Division, Washington, 1993.

[17] US Environmental Protection Agency, *Report to Congress on Cement Kiln Dust*, Office of Solid Waste, Washington, December 1993.

Table 2 Total discharge of TCDD/Fs from US pulp and paper mills in 1988 and 1993

	Effluent (g TEQ yr^{-1})	*Pulp* (g TEQ yr^{-1})	*Sludge* (g TEQ yr^{-1})	*Total* (g TEQ yr^{-1})
1988	352.8	496.8	338.4	1188.0
1993	28.8	32.4	36.0	97.2

Source: NCASI, 1994.

1.2×10^{-8} g TEQ kg^{-1} and 0.16×10^{-8} g TEQ kg^{-1} have been developed for kilns burning and not burning hazardous waste, respectively.[14] Multiplying the emission factors by the raw material throughput yields annual emission estimates of 210 g TEQ for kilns burning hazardous waste and 140 g TEQ for kilns not burning hazardous waste. Therefore, PCDD/F emissions to air from all cement kilns combined are estimated at about 350 g TEQ yr^{-1}.

Bleached Chemical Wood Pulp and Paper Mills

Pulp and paper mills are a much smaller source of PCDD/Fs in the environment than in the past. In 1988, the EPA and the US pulp and paper industry collaborated on the 104-Mill Study to measure levels of dioxins in effluent, sludge and pulp from 104 bleached chemical wood pulp and paper mills in the US. The study estimated the total discharge from all US pulp and paper mills to be 1188 g TEQ yr^{-1}.[18] In response to these findings, the pulp and paper industry expended more than $1 billion dollars to reduce TCDD and TCDF formation during bleaching operations.[19] The 1993 estimated discharge from all US pulp and paper mills is 97.2 g TEQ (Table 2).

Rigo[13] estimates that pulp and paper mills release 54–360 g TEQ yr^{-1} to the air and water (effluents in Table 2). Recent changes in pulp and paper mill operations have significantly reduced (but not eliminated) dioxin emissions, so the lower end of the range is probably more representative.

Motor Vehicles

Direct and indirect evidence indicates that motor vehicles are a source of environmental PCDD/Fs, with leaded gasoline emitting higher levels of PCDD/Fs into the environment than unleaded gasoline.[20-23] Leaded fuel-powered vehicles

---

[18] National Council of the Paper Industry for Air and Stream Improvement, *Progress in Reducing the TCDD/TCDF Content of Effluents, Pulps and Wastewater Treatment Sludges from the Manufacturing of Bleached Chemical Pulp*, special report, NCASI, New York, August 1994.

[19] American Forest and Paper Association, *The Paper Industry and Dioxin Reduction*, Washington, 1995.

[20] S. Marklund, C. Rappe, M. Tysklind and K. E. Egeback, *Chemosphere*, 1987, **16**, 29.

[21] A. G. Bingham, C. J. Edmunds, B. W. Graham and M. T. Jones, *Chemosphere*, 1989, **19**, 669.

[22] S. Marklund, R. Andersson, M. Tysklind, C. Rappe, K. E. Egeback, E. Bjorkman and V. Grigoriadis, *Chemosphere*, 1990, **20**, 553.

[23] H. Hagenmaier, V. Dawidowsky, U. B. Weber, O. Hutzinger, K. H. Schwind, H. Thoma, U. Essers, B. Buhler and R. Greiner, presented at Dioxin '90, 10th International Symposium on Chlorinated Dioxins and Related Compounds, Bayreuth, Federal Republic of Germany, 1990, short papers, vol. 2.

in the US were driven approximately 174 billion km in 1990 and are estimated to have emitted between 0.2 and 19 g TEQ, based on an emission factor range of 1.1–108 pg TEQ km^{-1} [14,21-23] In contrast, unleaded fuel-powered vehicles, which were driven approximately 3111 billion km in 1990, are estimated to have emitted 1.3 g TEQ,[14] based on an emission factor of 0.36 pg TEQ km^{-1}.[22]

Few data exist upon which to base an evaluation of the emissions of diesel-powered vehicles, but it is argued that the high average emission factor for heavy-duty diesel trucks, 5.4 ng TEQ kg^{-1},[24] makes these vehicles the predominant mobile source contributor of PCDD/PCDFs (Table 3).[25] Jones[25] estimates that total emissions of diesel-powered vehicles were 1105 g TEQ yr^{-1} in 1990 (Table 3) but predicts that by the year 2000 total emissions will drop to 545 g TEQ yr^{-1}, commensurate with the EPA's new diesel particulates control emissions standards. In contrast, the EPA[14] has estimated a much lower annual emission rate of 85 g TEQ yr^{-1} for diesel-powered vehicles in the US, using an emission factor of 0.5 ng TEQ km^{-1}.[22,26]

Residential Wood Burning

Chimney soot and bottom ash from wood-burning stoves and fireplaces in individual residences have yielded measurable levels of TCDD.[27,28] A study of PCDD/F emissions from residential wood burners in Switzerland found that a household stove burning natural beech wood yielded 0.77 ng TEQ kg^{-1} with the door open and 1.25 ng TEQ kg^{-1} with the door closed.[29] Applying an average emission factor[29,30] of 1 ng TEQ kg^{-1} for the 41.4 million metric tons of wood combusted in US homes each year[14] yields an estimate of 41 g TEQ yr^{-1} for emissions from residential wood burning.

Industrial Wood Burning

The common industrial practice of combusting wood to generate electricity and operate pulp and paper mills has been identified as a source of environmental PCDD/Fs. A study of PCDD/F emissions from seven wood-burning facilities in Switzerland showed that emissions ranged from 0.004 to 9.820 ng TEQ Nm^{-3} (0.048 to 117.8 ng TEQ kg^{-1}), with the majority of values being greater than 0.1 ng TEQ Nm^{-3} (the TEQ emission limit for waste incineration plants in Germany, Austria, the Netherlands and Sweden).[31] PCDD/F emissions from

[24] S. Larssen, E. M. Brevik and M. Oehme, presented at Dioxin '90, 10th International Symposium on Chlorinated Dioxins and Related Compounds, Bayreuth, Federal Republic of Germany, 1990, short papers, vol. 1.

[25] K. Jones, *J. Risk Anal.*, 1993, **13**, 245.

[26] M. Oehme, S. Larssen and E. M. Brevik, *Chemosphere*, 1991, **23**, 1699.

[27] R. E. Clement, H. M. Tosine and B. Ali, *Chemosphere*, 1985, **14**, 815.

[28] R. J. Wenning, M. A. Harris, D. J. Paustenbach and H. Bedbury, *Ecotoxicol. Environ. Saf.*, 1992, **23**, 133.

[29] B. Schatowitz, G. Brandt, F. Gafner, E. Schlumpf, R. Bühler, P. Hasler and T. Nussbaumer, presented at Dioxin '93, 13th International Symposium on Chlorinated Dioxins and Related Compounds, Vienna, Austria, September 1993.

[30] J. Vikelsoe, H. Madsen and K. Hansen, presented at Dioxin '93, 13th International Symposium on Chlorinated Dioxins and Related Compounds, Vienna, Austria, September 1993.

[31] J. Kolenda, H. Gass, M. Wilken, J. Jager and B. Zeschmar-Lahl, *Chemosphere*, 1994, **29**, 1927.

Table 3 Mobile source PCDD/PCDF emissions by vehicle classification[a]

| Vehicle class | Vehicle miles travelled (VMT)[b] and estimated emissions | | | |
| | 1990 | | 2000 | |
	VMT (km yr^{-1})	Emissions (g TEQ yr^{-1})	VMT (km yr^{-1})	Emissions (g TEQ yr^{-1})
Heavy-duty diesel truck	1.54×10^{11}	832	2.17×10^{11}	352
Medium-duty diesel truck	0.26×10^{11c}	140	0.35×10^{11}	57
Light-duty diesel truck	0.15×10^{11d}	4	0.20×10^{11}	2
Light-duty vehicle (noncat.)	2.42×10^{11e}	65	1.65×10^{11}	44
Light-duty vehicle (cat.)	21.8×10^{11}	59	31.4×10^{11}	85
Buses	0.09×10^{11f}	5	0.09×10^{11}	5
Totals		1105		545

Source: Jones (1993).
[a]Growth projections are based on 1980 to 1990 growth except for HDDT, which used LDV growth factor.
[b]VMT is from MVMA.
[c]Assumes 30% of all MDV in 1990 and 2000.
[d]Assumes 2% of all LDT in 1990 and 2000.
[e]Assumes 10% of LDV in 1990, 5% in 2000.
[f]Assumes all diesel buses in 1990 and 2000.

combustion of natural wood chips ranged from 0.79 to 2.57 ng TEQ kg^{-1}; emissions from combustion of waste wood chips ranged from 26 to 173.3 ng TEQ kg^{-1}.[29]

In 1990, an estimated 82.2 million metric tons of wood were burned in US industrial furnaces.[14] Assuming an average emission factor of 4.0 ng TEQ kg^{-1} wood from industrial wood-burning facilities in the US,[32,33] we estimate that 330 g TEQ are emitted each year from industrial furnaces. Using a slightly different emission factor, Rigo[13] estimated total emissions from wood burning (both residential and industrial) at 420 g TEQ yr^{-1}.

Summary

We estimate that 2100 g TEQ of PCDD/Fs are emitted into the US environment annually (Table 4). This compares with estimates of 9300 g TEQ by the EPA[14] and 4200 g TEQ by Rigo.[13] Municipal waste incinerators are the major contributor

---

[32] US Environmental Protection Agency, *National Dioxin Study Tier 4 – Combustion Sources. Engineering Analysis Report*, EPA-450/4-84-01h, Office of Air Quality Planning and Standards, Research Triangle Park, NC, 1987.
[33] California Air Resources Board, *Evaluation on a Woodwaste Fired Incinerator at Koppers Company, Orville, California*, Test Report No. C-88-065, Engineering Evaluation Branch Monitoring and Laboratory Division, CARB, May 29, 1990.

Table 4 Environmental emissions of PCDD/Fs from major identified sources

Sources	Travis and Nixon (1996) (g TEQ yr⁻¹)	EPA external Review draft (June 1994) (g TEQ yr⁻¹)	Rigo (1995) (g TEQ yr⁻¹)
Cement kilns	350	350	139
Hazardous waste incineration	35	35	100
Medical waste incineration	140	5100	325
Motor vehicles			1841
Leaded	10	10	
Unleaded	1	1	
Diesel	85	85	
Municipal waste incineration	1000	3000	1032
Pulp and paper mills	95	355	360
Wood-burning			420
Residential	40	40	
Industrial	320	320	
Total from known sources	2076	9300	4217

(48%) to known environmental sources of PCDD/Fs, followed by cement kilns, medical waste incinerators, wood combustion and pulp and paper mills.

5 Accumulation of PCDD/Fs in the Food Chain

The food chain has been shown to be the primary pathway of human exposure to PCDDs and PCDFs (Table 5).[34-39] We will examine the accumulation of dioxin in those foods found in the average American diet: fruits and vegetables, beef, milk and dairy products, and fish.

Accumulation in Fruits and Vegetables

Although studies on human exposure to dioxin typically focus on products of animal origin,[37-42] the accumulation of PCDD/Fs in vegetation is an important

[34] C. C. Travis and H. Hattemer-Frey, *Sci. Total Environ.*, 1991, **104**, 97.

[35] C. C. Travis and H. A. Hattemer-Frey, *Waste Manage.*, 1989, **9**, 151.

[36] S. Henry, G. Cramer, M. Bolger, J. Springer and R. Scheuplein, *Chemosphere*, 1992, **25**, 235.

[37] P. Fürst, C. Fürst and K. Wilmers, in *Biological Basis for Risk Assessment of Dioxins and Related Compounds, Banbury Report #35*, ed. M. A. Gallo, R. J. Scheuplein and K. van der Heijden, Cold Spring Harbor Laboratory Press, Plainview, NY, 1991.

[38] R. M. C. Theelen, in *Biological Basis for Risk Assessment of Dioxins and Related Compounds. Banbury Report #35*, ed. M. A. Gallo, R. J. Scheuplein and K. van der Heijden, Cold Spring Harbor Laboratory Press, Plainview, NY, 1991.

[39] H. Beck, K. Eckart, W. Mathar and R. Wittkowski, *Chemosphere*, 1989, **18**, 417.

[40] H. Beck, A. Dross and W. Mathar, *Chemosphere*, 1992, **25**, 1539.

[41] Ministry of Agriculture Fisheries and Food, *Dioxins in Food; Food Surveillance Paper*, No. 31, HMSO, London, 1992.

[42] B. Birmingham, B. Thorpe, R. Frank,. R. Clement, H. Tosine, G. Fleming, J. Ashman, J. Wheeler, B. D. Ripley and J. J. Ryan, *Chemosphere*, 1989, **19**, 507.

Table 5 US daily intake of food and PCDD/Fs (expressed in TEQs)

Food classification	Daily food intake	Daily PCDD/Fs intake (pg TEQ d^{-1})
Vegetable	218.7 g d^{-1} (wet weight)	6.6[a,c]
Vegetable oil	53.7 g d^{-1} (fat)	19.3[a]
Fruit	136.0 g d^{-1} (wet weight)	2.0[c]
Pork	8.3 g d^{-1} (fat)	2.3[a,b]
Beef	14.6 g d^{-1} (fat)	38[a,b]
Chicken	1.7 g d^{-1} (fat)	3.9[a,b]
Eggs	3.0 g d^{-1} (fat)	4.5[a,b,c]
Cow's milk	6.5 g d^{-1} (fat)	13.7[a]
Butter and cheese	10.5 g d^{-1} (fat)	22.1[a]
Fish and shellfish	0.34 g d^{-1} (fat)[d]	11.4[b]

[a]Based on average TEQ concentration from Theelen, 1991.
[b]Based on average TEQ concentration from Beck *et al.*, 1989.
[c]Based on average TEQ concentration from Beck *et al.*, 1994.
[d]Not including game fish.

first step in the environmental bioaccumulation and eventual transfer to humans.

Data on PCDD/Fs in US fruit and vegetable products are extremely limited. A Canadian study[42] found fruits, vegetables and wheat products to contain 0.004 pg TEQ g^{-1}, 0.002 pg TEQ g^{-1} and 0.0007 pg TEQ g^{-1}, respectively. While studies in other industrialized countries have found higher PCDD/F levels in vegetation – 0.015 pg TEQ g^{-1} in Germany[43] and 0.05 pg TEQ g^{-1} in the Netherlands[38] – all studies are in agreement that background contamination is well below 1 pg TEQ g^{-1}.

Assuming a PCDD/F concentration of 0.03 pg TEQ g^{-1} fresh weight for vegetables[38,43] and 0.015 pg TEQ g^{-1} fresh weight for fruit[44] and a daily consumption rate of 218.7 g for vegetables and 136.0 g for fruit,[45] we estimate that the US average daily intake of PCDD/Fs is 6.6 pg TEQ from vegetables and 2.0 pg TEQ from fruit (Table 5).

Accumulation in Beef and Milk

Farm animals are exposed to PCDD/F *via* the ingestion of contaminated forage, grains and soil. Owing to its high lipophilicity, dioxin readily bioaccumulates in cattle and has been detected in beef, milk and dairy products. Several studies point out that meat and dairy products account for up to 60% of the total daily human intake of PCDD/Fs in industrialized nations.[39,46] PCDD/F levels in beef

[43] H. Beck, A. Dross and W. Mathar, *Environ. Health Perspect.*, 1994, **102**, 173.
[44] R. Schroll and I. Scheunert, *Chemosphere*, 1993, **26**, 1631.
[45] J. J. Putnam and J. E. Allshouse, *Food Consumption, Prices, and Expenditure, 1970–90*, Statistical Bulletin No. 840, Commodity Economics Division, Economic Research Service, US Department of Agriculture, Washington, August 1992.
[46] P. Fürst, C. Fürst and K. Groebel, *Chemosphere*, 1990, **20**, 787.

have been measured at: (1) 2.6 pg TEQ g^{-1} fat in Germany,[39] (2) 1.75 pg TEQ g^{-1} fat in the Netherlands[47] and (3) 2.9 pg TEQ g^{-1} fat in Canada.[42] Assuming a PCDD/F concentration of 2.6 pg TEQ g^{-1} beef fat[38,39] and a daily consumption rate of 14.6 g beef fat,[45] we estimate the US average daily intake of PCDD/Fs from beef to be 38 pg TEQ (Table 5).

The high lipid content of cow's milk, cheese and butter make dairy products an important pathway of exposure to PCDD/F. Although one study[46] of milk and milk products indicated PCDD/F levels between 0.76 and 2.62 pg TEQ g^{-1} fat, concentrations have been consistently measured in the narrower range of 1.8–2.1 pg TEQ g^{-1} fat.[38,39,43] Cheese exhibits similar contamination levels on a lipid basis,[38] but butter has shown less consistency. One study[39] measured PCDD/F concentrations in butter at 0.8 pg TEQ g^{-1} fat, considerably below those in cow's milk. Assuming a PCDD/F concentration of 2.1 pg TEQ g^{-1} fat for dairy products[38,48] and an average daily intake of 6.5 g fat d^{-1} for cow's milk and 10.5 g fat d^{-1} for butter and cheese,[45] we estimate the US daily intake of PCDD/Fs to be 13.7 pg TEQ from milk and 22.1 pg TEQ from butter and cheese (Table 5).

Is Milk Packaging a Major Pathway of Human Exposure to Dioxin?

Packaging material was examined as a potential contamination source for food after migration of PCDD/Fs from chlorine-bleached cartons into milk had been demonstrated.[49] Other studies confirmed that PCDD/Fs carried over from bleached milk cartons and bleached coffee filters.[50,51] However, the paper industry recently agreed to reduce the amount of PCDDs and PCDFs in bleached paper products to below 1 pg TEQ g^{-1} (ppt); therefore, migration of PCDD/Fs from milk packaging material is no longer considered a source of concern.

Accumulation in Fish

PCDD/F levels measured in fish from lakes and rivers in the US confirm that low-level contamination of fish is widespread.[52] The average concentration of PCDD/Fs in North American fish is 1.16 pg TEQ g^{-1} whole weight.[14] However, the level of contamination is highly dependent on the origin of the sample, the age of the specimen and the season when it was caught.

Travis and Hattemer-Frey[53] reviewed EPA data on 2,3,7,8-TCDD concentrations in fish collected from 304 urban sites across the US, including the Great Lakes

[47] R. M. C. Theelen, A. K. D. Liem, W. Slob and J. H. van Wijnen, *Chemosphere*, 1993, **27**, 1625.

[48] A. P. J. M. de Jong, A. Liem, R. Hoogerbrugge, R. Theelen and H. A. van't Klooster, in *National Institute of Public Health and Environmental Protection*, Report No. 730501.002, Brussels, 1990.

[49] J. J. Ryan, L. G. Panopio, D. A. Lewis and D. F. Weber, *J. Agric. Food Chem.*, 1991, **39**, 218.

[50] H. Beck, A. Dross, W. Mathar and R. Wittkowski, *Chemosphere*, 1990, **21**, 789.

[51] H. Beck, A. Dross, K. Eckart, W. Mathar and R. Wittkowski, *Chemosphere*, 1989, **19**, 655.

[52] US Environmental Protection Agency, *The National Dioxin Study: Tiers 3,4,5, and 7*, EPA 440/4-87-003, Office of Water Regulation and Standards, Monitoring and Data Support Group, Washington, 1987.

[53] C. C. Travis and H. Hattemer-Frey, *Environ. Int.*, 1990, **16**, 155.

	No fish consumption	*Normal fish consumption*	*High fish consumption*
2,3,7,8-TCDD	1.8	2.5	8.0
TEQ	17.5	25.8	63.5

Source: Rappe (1992).

region,[52] and those in fish taken from sites downstream of pulp and paper mills.[54] The concentration of TCDD in fish fillets from sites near pulp and paper mills was 0.9 pg g^{-1}, which is three times higher than urban fish fillet concentrations (0.3 pg g^{-1}) but about two times lower than the TCDD concentration of fish fillets from the Great Lakes region (2.3 pg g^{-1}).

An analysis of five fish samples obtained from a supermarket in upstate New York yielded these measurements of PCDD/Fs: 0.03 pg TEQ g^{-1} for haddock, 0.02 pg TEQ g^{-1} for haddock fillet, 0.13 pg TEQ g^{-1} for crunchy haddock, 0.023 pg TEQ g^{-1} for perch and 0.023 pg TEQ g^{-1} for cod, all on a wet weight basis.[6]

Assuming that fish and shellfish (not including game fish) contain an average PCDD/F concentration of 33.6 pg TEQ g^{-1} fat[43] and that average daily consumption of fat from fish and shellfish is 0.34 g,[45] we estimate that the US average daily intake of PCDD/F from fish and shellfish is 11.4 pg TEQ d^{-1} (Table 5).

Is Fish Ingestion a Major Pathway of Human Exposure to Dioxin?

We estimate that ingestion of fish contributes about 6–9% to total human PCDD/F intake in the US. However, it can contribute up to 30% in countries with high fish consumption. A Swedish study[55] found that people with a high fish consumption had a body burden (measured in TEQs) three times higher than that of people with normal fish consumption (see Table 6).

6 Comparing Source Emissions to Deposition Estimates

Studies comparing emissions from known sources of PCDD/Fs to aerial deposition rates in Sweden[56] and the United Kingdom[57,58] found that annual depositions were 8–20 times higher than annual emissions, suggesting that the bulk of dioxin present in the atmosphere is due to unidentified sources. Estimates of atmospheric deposition rates for PCDD/Fs vary from 1–2 ng TEQ m^{-2} yr^{-1} for limited US data to 2–30 ng TEQ m^{-2} yr^{-1} for more extensive European data. Based on an assumed combined wet and dry particulate and gaseous deposition

---

[54] US Environmental Protection Agency, *Dioxin Levels in Fish Near Pulp and Paper Mills*, Office of Water Regulation and Standards, Monitoring and Data Support Group, Washington, 1988.

[55] C. Rappe, *Chemosphere*, 1992, **25**, 231.

[56] C. Rappe, in *Biological Basis for Risk Assessment of Dioxins and Related Compounds. Banbury Report #35*, ed. M. A. Gallo, R. J. Scheuplein and K. van der Heijden, Cold Spring Harbor Laboratory Press, Plainview, NY, 1991, p. 121.

[57] S. J. Harrad, A. P. Stewart and K. C. Jones, presented at Dioxin '92, 12th International Symposium on Chlorinated Dioxins and Related Compounds, Tampere, Finland, August 1992.

[58] S. J. Harrad and K. C. Jones, *Sci. Total Environ.*, 1992, **126**, 89.

in the continental US of 9.2 ng m^{-2} yr^{-1} for urban areas (2% of total area) and 4.4 ng m^{-2} yr^{-1} for rural areas (98% of total area), Rigo[13] has estimated that the annual average atmospheric deposition of dioxin in the US is 33 000 g TEQ yr^{-1}. Rigo's estimate falls within the range of 20 000–50 000 g TEQ yr^{-1} suggested by the EPA.[14]

The above estimates suggest that there may be significant unidentified sources of PCDD/Fs. This possibility is highlighted by Rigo,[13] who assigns 82% of total annual emissions in the US to 'other, unidentified and natural' sources, and by Travis and Hattemer-Frey,[34] who have estimated that as much as 90% of total TCDD input into the US environment is unaccounted for by specific sources. This view is questioned by Thomas and Spiro,[59] who believe that careful analysis of dioxin emissions provides no evidence for significant missing dioxin sources.

7 Human Exposure to PCDDs and PCDFs

For risk assessment purposes, an important objective in evaluating the environmental fate of PCDD/Fs is predicting the major pathways of human exposure. It is well established that the food chain, especially meat and dairy products, accounts for more than 90% of human exposure to PCDD/Fs and perhaps as much as 99% of human exposure to 2,3,7,8-TCDD.[34] In industrialized countries, the average daily intake *via* food (the major route of exposure to dioxins and furans) ranges from 1.5 to 2.5 pg TEQ kg^{-1} body weight.

Other potential pathways of exposure – air inhalation, ingestion of water and soil, and dermal contact with soil – are much less significant to total daily intake of dioxins. PCDD/F intake *via* air inhalation, water ingestion and soil ingestion is estimated to be 3.2 pg TEQ d^{-1} (2–3% of total daily intake).[38] Dermal contact with soil is negligible, however, constituting an estimated daily intake of 0.15 pg TEQ d^{-1}.[38]

Ultimate proof of the widespread exposure to TCDD is exemplified by the fact that virtually all human adipose tissue samples contain TCDD at trace levels of 6 ppt.[60] Adipose tissue of non-occupationally exposed humans has been found to contain concentrations of 18–122 pg TEQ g^{-1} fat, with a mean value of 56 pg TEQ g^{-1} fat,[43] a value 22 times higher than the concentration in beef fat (2.6 pg TEQ g^{-1} fat).

Dietary exposure to dioxin may be increased or decreased significantly by special consumption habits. For example, populations with high fish consumption can markedly exceed the international estimated daily intake of 2.3 pg TEQ kg^{-1} body weight. In contrast, a vegetarian diet may reduce intake of PCDD/PCDFs by 98%.[6] Nevertheless, Welge *et al.* compared blood levels of PCDD/Fs in 24 vegetarians and 24 omnivores and found no significant difference (vegetarian mean = 32.6 pg TEQ g^{-1} lipid; omnivore mean = 34.3 pg TEQ g^{-1} lipid) between the two groups.[61] Given the high levels of PCDD/Fs in dairy products and the fact that such foodstuffs often constitute a disproportionately large component of the diet of vegetarians, this is unsurprising.

[59] V. M. Thomas and T. G. Spiro, *Environ. Sci. Technol.*, 1996, **30**, 82A.
[60] M. Gough, *J. Toxicol. Environ. Health*, 1991, **104**, 129.
[61] P. Welge *et al.*, *Organohalogen Compd.*, 1993, **13**, 13.

Table 7 Estimated exposure to PCDD/Fs (pg TEQ d^{-1})

Route/source	The Netherlands[a]	UK[b]	Germany[c]	US[d]	North America[e]
Inhaled air	2			2	2.2
Ingested air	1			1	
Water					
Soil via skin	0.2			0.2	
Soil ingested	0.1			0.1	0.8
Total air, soil, water	3.3 (3%)			3.3 (3%)	3.0 (3%)
Vegetables	1.8-7	15	3.7	6.6	
Vegetable oil	14	19	0.4	19	
Fruits			2.0	2	
Cereals		5.3			
Total vegetable	18.5 (16%)	39.3 (31%)	6.1 (3%)	27.6 (22%)	
Pork	4.2		35.9 (includes all meats)	2.3	12.2
Beef	13			38	37.0
Chicken	4.8 (with eggs)	5.6		3.9	12.9
Eggs		4.6	5.9	4.5	4.1
total meat and eggs	22 (19%)	42.2 (34%)	41.8 (26%)	48.7 (38%)	66.2 (56%)
Cow's milk	17	23		13.7	17.6
Cheese and butter	26	12		22.1	24.1
Total dairy product	43 (36%)	35 (28%)	55.1 (34%)	35.8 (28%)	41.7 (35%)
Seawater fish	14				
Freshwater fish	10				
Fish oil	7.2				
Total fish	31.2 (26%)	7.7 (6%)	60.5 (37%)	11.4 (9%)	7.8 (6%)
Total intake	118	125	164	127	119

[a]Theelen (1991).
[b]Ministry of Agriculture (1992).
[c]Beck *et al.* (1994).
[d]Travis and Nixon (1996).
[e]EPA (1994).

29

Estimated Daily Intake

Estimates for total daily intake of dioxin *via* food in industrialized nations are fairly consistent. The average background daily intake of PCDD/Fs by the general population of (1) the US is 1.8 pg TEQ kg^{-1} body weight (our estimate), (2) Germany is 1.3–2.3 pg TEQ kg^{-1} body weight,[43] (3) Japan is 1.2 pg TEQ kg^{-1} body weight[62] and (4) Canada is 1.5 pg TEQ kg^{-1} body weight.[42] The mean background dietary intake of PCDD/Fs in industrialized countries is about 133 pg TEQs d^{-1} (for an individual weighing 70 kg) (Table 7), with 2,3,7,8-TCDD accounting for about 20% of the total daily intake of PCDD/Fs (expressed in TEQs).

Health Standards

The EPA has established[63] a virtually safe dose of 0.006 pg TEQ kg^{-1} d^{-1}, which represents one excess cancer in a million in exposure population. Many European countries use the World Health Organization value of 10 pg kg^{-1} d^{-1} Tolerable Daily Intake (TDI). US daily intake of PCDD/Fs by the general population (1.8 pg TEQ kg^{-1} d^{-1}) greatly exceeds the EPA's virtually safe dose, but falls within the World Health Organization TDI.

Conclusions

Ambient measurements confirm that environmental PCDD/F contamination is widespread and that virtually everyone – regardless of age, gender, or geographic location – is exposed to these compounds on a daily basis. We identify MSW incinerators as the major known source (48%) of environmental PCDD/Fs, followed by cement kilns, medical waste incinerators, wood combustion and pulp and paper mills. However, these known sources appear to account for only a fraction (10–30%) of the total annual atmospheric deposition of PCDD/Fs in the US. Thus, either significant unidentified sources of dioxin exist or most dioxin presently circulating in the atmosphere results from re-suspension and/or volatilization of historically deposited dioxin.

The food chain is the primary pathway of human exposure to dioxin, with meat (38%) and dairy products (28%) dominating. Fish ingestion can be a significant contributor in countries with high fish consumption (26% of total intake in the Netherlands), but is not an important factor in the US. The exact contribution of fruits and vegetables is unclear, but vegetable oil does appear to play a role in human dioxin exposure. Inhalation and consumption of contaminated water and soil are not major sources of human exposure to TCDD.

Estimates of total daily intake of dioxin *via* food in industrialized nations are fairly consistent, ranging from 1.3 to 2.3 pg TEQ kg^{-1} d^{-1}.[14,42,43,62] The average long-term intake of PCDD/Fs by the general population of the US is about 127 pg TEQ d^{-1}, with an intake of TCDD of about 27 pg d^{-1}.

[62] M. Ono, Y. Kashima, T. Wakimoto and R. Tatsukawa, *Chemosphere*, 1987, **16**, 1823.
[63] US Environmental Protection Agency, *Ambient Water Quality Criteria for 2,3,7,8-Tetrachlorodibenzo-p-dioxin*, EPA-440/5-84-007, EPA, Washington, 1984.

Biological Uptake and Transfer of Polychlorinated Dibenzo-*p*-dioxins and Dibenzofurans

MICHAEL S. McLACHLAN

1 Introduction

Polychlorinated dibenzo-*p*-dioxins and dibenzofurans (PCDD/Fs) have over the last 15 years been the subject of intense environmental chemistry research. This interest is rooted in the persistence and hydrophobicity of these compounds, which impart to them a strong tendency to bioaccumulate. This, coupled with the fact that a number of PCDD/Fs have been found to be toxic at very low concentrations, has fuelled interest in understanding their behaviour in the environment. One crucial component of the environmental behaviour is the biological uptake and transfer of PCDD/Fs, since this determines the exposure of biota to these compounds.

There are 135 chlorinated dibenzofuran and 75 dibenzo-*p*-dioxin congeners. The research to date has concentrated on the tetra- to octachlorinated compounds, since PCDD/Fs with less than four chlorine atoms are not found in human tissue. Of particular toxicological relevance are the 2,3,7,8-substituted congeners, of which there are 17. In order to simplify the interpretation of PCDD/F data, a system known as toxicity equivalents was developed to express the toxicity of all PCDD/Fs in a sample in terms of the most toxic member of this family, 2,3,7,8-Cl_4DD.[1]

A good deal of the concern surrounding this class of compounds has arisen from the proximity of current background levels of human exposure to the exposure levels at which toxic effects are suspected. It has been demonstrated that this background exposure is due almost exclusively to the ingestion of animal fat, the average uptake in Europe and North America being approximately equally divided between fish, meat and dairy products (see Table 1).[2-5] Owing to the

[1] F. W. Kutz, D. G. Barnes, D. P. Bottimore, H. Greim and E. W. Bretthauer, *Chemosphere*, 1990, **20**, 751.
[2] P. Fürst, C. Fürst and W. Groebel, *Chemosphere*, 1990, **20**, 787.
[3] R. M. C. Theelen, A. K. D. Liem, W. Slob and J. H. van Wijnen, *Chemosphere*, 1993, **27**, 1625.
[4] Ministry of Agriculture, Fisheries and Food, *Dioxins in Food; Food Surveillance Paper No. 31* HMSO, London, 1992.
[5] B. Birmingham, A. Gilman, D. Grant, J. Salminen, M. Boddington, *et al.*, *Chemosphere*, 1989, **19**, 637.

Country	Milk products	Meat/eggs	Fish products	Other	Source
Germany	31.5	32.2	31.7	4.5	Ref 2
The Netherlands	42.6	24.6	31.1	1.7	Ref 3
Great Britain	28	34.3	21.4	16.2	Ref 4
Canada	22.2	43.7	12.1	21.9	Ref 5

Table 1 Sources of human exposure to PCDD/Fs (in %)

pre-eminent importance of agricultural products for human exposure, this paper will focus on bioaccumulation of PCDD/Fs in agricultural food chains. A large fraction of the PCDD/F exposure through meat is due to beef consumption, and beef and dairy cattle together account for about one half of human exposure. Cattle for their part obtain most of their PCDD/Fs through grass.[6] Hence the grass–cattle–milk/beef pathway is critical for human exposure and will receive special emphasis.

This chapter begins at the bottom of the agricultural food chain, looking at the uptake of PCDD/Fs in plants from soil. The second section deals with the uptake in plants from the atmosphere. The attention then switches to the next link in the food chain, and the transfer from plants and soil to livestock and animal food products is examined. This is followed by a discussion of uptake in humans from food and in infants from mothers' milk. The chapter concludes with a short summary, an overall perspective on bioaccumulation in the agricultural food chain, and a brief reference to some consequences of this knowledge for risk assessment.

2 Soil/Plant Transfer

There are two possible sources of PCDD/Fs to vegetation: the atmosphere and soil. Initially it was thought that PCDD/Fs would not be present in the atmosphere in quantities sufficient to contaminate plants owing to their low volatility, and early research in this area focused on uptake from soil. There are three possible pathways of soil-bound PCDD/Fs to aerial plant parts: root uptake and translocation, volatilization followed by adsorption to foliage, and transfer of soil particles (see Figure 1). The first of these pathways has received the most attention.

Root Uptake and Translocation

An early paper by Isensee and Jones[7] concluded, on the basis of experiments with radioactively labelled $2,3,7,8-Cl_4DD$, that uptake from soil and translocation to aerial plant parts was unlikely. However, when elevated levels of $2,3,7,8-Cl_4DD$ were found in vegetation from the Seveso area that had begun growing one year after an accident which released large quantities of this compound into the environment, this was interpreted as evidence of uptake and translocation from the contaminated soil.[8] A second group of researchers analysed vegetation samples from the same site and came to the opposite conclusion, suggesting that

[6] M. S. McLachlan and O. Hutzinger, *Organohalogen Compd.*, 1990, **1**, 479.
[7] A. R. Isensee and G. E. Jones, *J. Agric. Food Chem.*, 1971, **19**, 1210.
[8] S. Cocucci, F. Di Gerolamo, A. Verderio, A. Cavallaro, G. Colli, *et al.*, *Experientia*, 1979, **35**, 482.

Figure 1 Pathways of
PCDD/Fs to plants

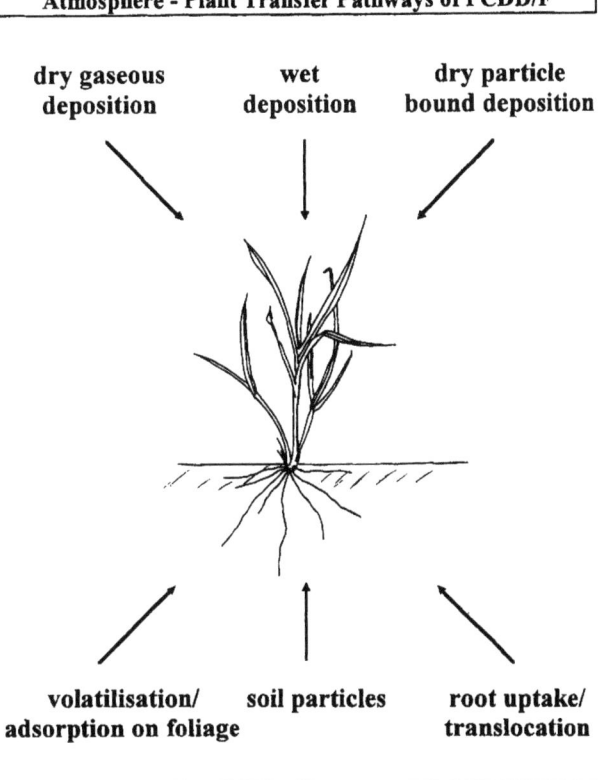

the contamination with 2,3,7,8-Cl$_4$DD was due to atmospheric deposition and not root uptake/translocation.[9] Two research groups then conducted laboratory experiments growing bean and maize plants in media containing 2,3,7,8-Cl$_4$DD. While the first group identified vaporization from soil and subsequent deposition on leaves as one transport pathway, they concluded that root uptake/translocation was a much more important mechanism.[10] The second group observed that the leaf concentrations did not increase with time or soil concentration when the plants were grown in the same atmosphere and concluded that root uptake/translocation was less significant than volatilization and subsequent deposition.[11] This highly contradictory literature has been reviewed in detail.[12]

A major reason for the contradictions in this early literature is that in all experiments both root uptake/translocation as well as volatilization from soil followed by deposition to the leaf surface were possible. The authors used

[9] H.-K. Wipf, E. Homberger, N. Neuner, U. B. Ranalder, W. Vetter, *et al.*, in *Chlorinated Dioxins and Related Compounds: Impact on the Environment*, ed. O. Hutzinger *et al.*, Pergamon Press, New York, 1982, p. 115.

[10] G. A. Sacchi, P. Vigano, G. Fortunati and S. M. Cocucci, *Experientia*, 1986, **42**, 586.

[11] S. Facchetti, A. Balasso, C. Fichtner, G. Frare, A. Leoni, *et al.*, *Chemosphere*, 1986, **15**, 1387.

[12] G. A. Kew, J. L. Schaum, P. White and T. T. Evans, *Chemosphere*, 1989, **18**, 1313.

circumstantial evidence to evaluate the relative importance of the two pathways and came to different conclusions. This problem was finally overcome with an experimental apparatus which successfully excluded the volatilization/deposition pathway.[13] Also working with bean and maize, the authors clearly showed that uptake from soil followed by translocation to aerial plant parts was negligible. This result is in agreement with what one would expect from very hydrophobic compounds such as the PCDD/Fs on the basis of current understanding of the transport of xenobiotics in plants.[14] It is also consistent with the subsequent literature, in which all authors come to the conclusion that the soil uptake/translocation pathway is insignificant[15-20].

An exception to this rule has, however, been discovered: plants of the genus Cucurbita readily take up and translocate PCDD/Fs from the soil to the leaves and fruit.[21] It is suspected that these plants produce root exudates which form a complex with PCDD/Fs that is then reabsorbed and transported in the xylem throughout the whole plant.[22,23]

Volatilization from Soil/Adsorption to Foliage

Having shown that root uptake/translocation is an insignificant pathway, the implication of many of the laboratory studies cited above is that volatilization from soil and subsequent adsorption on aerial plant parts is an important uptake mechanism. However, using a mathematical model of this process it was shown that in the field only the lowest few centimetres of plant material would be contaminated if volatilization occurred.[24]

This model also indicated that the degree of transfer to aerial plant parts should decrease with increasing degree of chlorination (decreasing vapour pressure) of the compound.[24] This preferential transfer of the lower chlorinated congeners was demonstrated in an experiment growing grass on contaminated soil in a chamber with a low ventilation rate.[25] However, in those cases where aerial parts of plants grown under natural conditions on highly contaminated soils have been analysed, the pattern of PCDD/Fs found in the plants has been similar to the pattern present in the soil.[16,18] This indicates that the volatilization/adsorption pathway is not significant, since in this case the pattern

[13] J. K. McCrady, C. McFarlane and L. K. Gander, *Chemosphere*, 1990, **21**, 359.

[14] G. G. Briggs, R. H. Bromilow and A. A. Evans, *Pestic. Sci.*, 1982, **13**, 495.

[15] R. Schroll and I. Scheunert, *Chemosphere*, 1993, **26**, 1631.

[16] A. Hülster and H. Marschner, *Chemosphere*, 1993, **27**, 439.

[17] J. F. Müller, A. Hülster, O. Päpke, M. Ball and H. Marschner, *Chemosphere*, 1993, **27**, 195.

[18] G. H. M. Krause, T. Delschen, P. Fürst and D. Hein, *UWSF - Z. Umweltchem. Okotox.*, 1993, **5**, 194.

[19] K. Welsch-Pausch, M. S. McLachlan and G. Umlauf, *Environ. Sci. Technol.*, 1995, **29**, 1090.

[20] G. H. M. Krause, B. Prinz and L. Radermacher, in *Kriterien zur Beurteilung organischer Bodenkontaminationen: Dioxine (PCDD/F) und Phthalate*, ed. G. Kreysa and J. Wiesner, Dechema, Frankfurt a. M., 1995, p. 287.

[21] A. Hülster, J. F. Müller and H. Marschner, *Environ. Sci. Technol.*, 1994, **28**, 1110.

[22] A. Hülster and H. Marschner, *Organohalogen Compd.*, 1994, **20**, 31.

[23] A. Hülster and H. Marschner, *Organohalogen Compd.*, 1995, **24**, 493.

[24] S. Trapp and M. Matthies, *UWSF - Z. Umweltchem. Okotox.*, 1994, **6**, 157.

[25] G. H. M. Krause, in *Aus der Tätigkeit der LIS 1992*, Landesanstalt für Immissionsschutz NRW, Essen, 1993, p. 69.

in the plant would have had a much stronger contribution from the lower chlorinated homologues than the pattern in the soil.

Transfer of Soil Particles to Foliage

Having excluded soil uptake/translocation and volatilization/adsorption, it can be concluded that the direct transfer of soil particles is the primary pathway of PCDD/Fs in soil to aerial plant parts under environmental conditions. This mechanism is consistent with the similar homologue patterns found in highly contaminated soils and plants grown on them. The degree of transfer of soil particles is a function of many variables, including the plant species, the soil properties and the climate. This transfer is often described using transfer factors defined as the quotient of the concentrations in the vegetation and the soil, both on a dry weight basis. Transfer factors of less than 0.001 were measured for PCDD/Fs in lettuce, potato shoots and hay grown on contaminated soil.[16] Transfer factors of less than 0.01 were reported for hay and garden vegetables from a wide range of different sites, whereby the authors make the point that their estimates are upper limits since the concentrations in the vegetation were not corrected for the input from aerial deposition.[20] On the basis of these results it can be concluded that the soil/vegetation transfer factor is very low for garden vegetables.

This is also true for root vegetables. In field studies it has been found that the transfer factors for root vegetables are similar to or even lower than those for aerial vegetation.[16,26] It has also been demonstrated that the PCDD/Fs remain primarily on the surface, not penetrating into root vegetables and fruit.[16,17] Since the PCDD/Fs are transferred *via* soil particles, washing the vegetables serves to reduce the PCDD/F concentrations.[16]

Hay and other fodder crops are more subject to soil contamination. In addition to natural processes like wind and splash, fodder crops are often contaminated with soil during harvesting. Since most crops such as hay are not washed prior to feeding, livestock may ingest considerable quantities of soil. For instance, it was estimated that in the Netherlands the soil content of grass silage was several percent.[27] This is much higher than the $<0.1\%$ reported for manually harvested grass and vegetables cited above,[16] and was calculated to result in an average soil ingestion rate of 400 g d^{-1} for milk cows during winter feeding. Root crops such as fodder beets can be an even more significant source of soil to livestock. The degree of feed contamination will again be a function of many variables including soil, climate, crop and harvesting technique. There has been little study in this area to date, although this is the most important vector of soil-bound PCDD/Fs into the agricultural food chain.

Summary

In summary, root uptake/translocation is an insignificant pathway of PCDD/Fs to aerial plant parts, the only known exceptions being several members of the

[26] B. Prinz, G. H. M. Krause and L. Radermacher, *Chemosphere*, 1991, **23**, 1743.
[27] P. L. M. Berende, *Grondopname door melkkoeien*, Institute for Livestock Feeding and Nutrition Research, Internal Report No. 312, Lelystad, 1990.

genus Cucurbita. PCDD/Fs adsorb to the surfaces of root vegetables, but only a very small fraction is transported deeper into the root tissue. Volatilization followed by deposition to aerial leaf parts is also of secondary importance under natural conditions. The main pathway of PCDD/Fs in soil into foliage is the direct transport of soil particles to the surface of the leaves and fruits. The soil/plant transfer rates are very low, estimated to be less than 0.001, and much of this contamination can be removed by cleaning. Of more concern is the contamination of animal feed with soil, both because the harvesting of feed crops often is accompanied by the entrainment of considerable quantities of soil, and because many feed crops are not washed prior to feeding. However, as the next section shows, all soil-related sources of PCDD/Fs to the agricultural food chain are small compared to the atmospheric sources.

3 Atmosphere/Plant Transfer

As a consequence of the preoccupation of the dioxin research community with soil as a source of plant contamination, the investigation of atmospheric deposition began at a late date. In 1990 it was reported that the concentrations of PCDD/Fs in grass and in soil growing in a rural area in Central Europe were similar. If soil was the source of the PCDD/Fs in the grass, this would imply that the transfer rate was about 1. This is implausible, and the authors concluded that atmospheric deposition was the source.[6] In two other studies, no correlation was found between the PCDD/F levels in grass and garden vegetables and those in soil measured at various locations in northwestern Germany, and the authors came to the same conclusion.[20,28] When care was taken to prevent transfer of soil particles to crops growing on highly contaminated soil in an area with low air concentrations, the air and not the soil homologue pattern was found in the aerial plant parts. It was concluded that atmospheric deposition was the primary source, even when the soil concentrations were very high.[16]

The Relative Contributions of the Different Forms of Atmospheric Deposition

Atmospheric deposition of PCDD/Fs can be divided into three different forms: dry gaseous deposition, dry particle-bound deposition and wet deposition (see Figure 1). Dry gaseous deposition is the diffusion of gaseous chemicals from the atmosphere to the plant surface. Dry particle-bound deposition occurs when particulate matter that contains the contaminant is deposited on the plant surface. Wet deposition can transport chemicals either in dissolved form, in particles trapped in the precipitation, or sorbed to the surface of water droplets or ice crystals. There are a multitude of forms of wet deposition, ranging from hail through rain to fog and dew fall.

The manner in which a given compound is deposited is determined by the form in which it is present in the atmosphere. Semi-volatile organic compounds partition between the gas and particle phases. The lower the vapour pressure of a

[28] A. Hembrock-Heger, *Organohalogen Compd.*, 1990, **1**, 475.

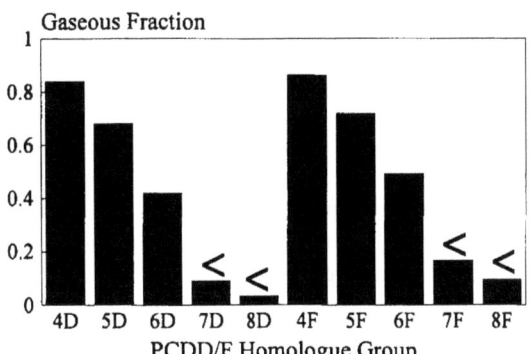

Figure 2 Average fraction of the total air concentration of the PCDD/F homologues that is present in the gas phase during the summer months (Data from ref. 30)

compound, the greater is its tendency to partition onto particles and hence the greater its likelihood to reach the plant in the form of dry or wet particle-bound deposition. Similarly, the importance of wet deposition of dissolved chemical is dependent on its air/water partitioning behaviour. The PCDD/Fs have low vapour pressures and only a moderate tendency to partition from air into water, and hence wet deposition of dissolved chemical is of little importance compared to particle-bound wet deposition. The gas/particle partitioning behaviour during the growing season is very variable, ranging from almost completely gaseous for the more volatile Cl_4DF to virtually completely particle bound for Cl_8DD (see Figure 2).[29,30] This makes the determination of the predominant form of deposition particularly interesting and challenging, since the Cl_4DF would be expected to be subject to dry gaseous deposition, while in the case of Cl_8DD wet and dry particle-bound deposition might be expected to dominate.

The first attempts to model PCDD/F accumulation in plants treated only dry particle-bound deposition.[31,32] However, as the research community became aware of the fact that a large fraction of the toxicologically important lower chlorinated PCDD/Fs were present in the gas phase during the growing season, the need to treat other forms of deposition was recognized.[6,33] This led to a first attempt to model all three forms of uptake of 2,3,7,8-Cl_4DD in grass to see if any of the three forms of atmospheric deposition could be ruled out.[34] This modelling effort revealed the limits of current understanding of the deposition processes, and the resulting model predictions were at best order-of-magnitude estimates. It was found that each of the deposition pathways could theoretically by itself explain the grass concentrations measured in the environment.

An experiment was designed to establish the relative contributions of each of the different forms of deposition. A system of greenhouses with different exposure environments was used together with adjacent outdoor plots to study the uptake of PCDD/Fs in ryegrass. Under the experimental conditions (summer and early autumn at a semi-rural location in Central Europe) it was found that the uptake

[29] M. S. McLachlan and O. Hutzinger, *Organohalogen Compd.*, 1990, **1**, 441.

[30] M. Hippelein, H. Kaupp, G. Dörr, M. McLachlan and O. Hutzinger, *Chemosphere*, 1996, **32**, 1605.

[31] P. Connett and T. Webster, *Chemosphere*, 1987, **16**, 2079.

[32] J. B. Stevens and E. N. Gerbec, *Risk Anal.*, 1988, **8**, 329.

[33] A. Reischl, M. Reissinger, H. Thoma and O. Hutzinger, *Chemosphere*, 1989, **19**, 467.

[34] M. S. McLachlan, *Organohalogen Compd.*, 1991, **6**, 183.

of the Cl_4–Cl_6DD/Fs in the ryegrass was primarily due to dry gaseous deposition. The deposition of small particles was not an important pathway for any of the compounds. No firm conclusions could be drawn for the more strongly particle associated Cl_7–Cl_8DD/Fs, but there were some suggestions that the dry deposition of large particles or wet deposition played an important role.[19,35]

This conclusion was supported by data from field studies. In comparing homologue patterns in vegetation and in the gaseous and particle phases of the atmosphere, it was observed that the relative contribution of the lower chlorinated PCDD/Fs was much higher in vegetation than in particles. It would only be possible to attribute this to particle-bound deposition if the higher chlorinated congeners were degraded on the plant surface or if there was a preferential transfer of lower chlorinated PCDD/Fs from the particles to the plant followed by erosion of the particles. These explanations were thought to be unlikely, and the authors concluded that gaseous deposition contributed a large part of the plant uptake of the Cl_4–Cl_6DD/Fs.[6,36]

A different interpretation of field data was presented in a third study.[26] Vegetation and bulk deposition samples were collected from several locations in and around the Ruhr industrial region in Germany and analysed. A good correlation between the PCDD/F concentrations (expressed in toxicity equivalents (TEQs)) in vegetation and in bulk deposition was observed. On the strength of this correlation it was concluded that particle-bound deposition is the main pathway of PCDD/Fs to vegetation. One explanation for the different conclusion in this study could be that there are more large particles in the air in the Ruhr area. This would be expected close to combustion sources. Large particles are deposited quickly and could thus result in a larger contribution of particle-bound deposition in the Ruhr area than in areas more remote from sources where there are few large particles in the atmosphere. On the other hand, the data interpretation in this study does not demonstrate causality. The levels of PCDD/Fs in bulk deposition and in the gas phase will be correlated, since an increase in the gaseous concentrations results in an increase in the particle-bound concentrations through the partitioning phenomenon, hence increasing the bulk deposition. While a good correlation was obtained between plant levels and bulk deposition, there is likely an equally good correlation between plant levels and the gaseous concentrations. The authors could have supported their hypothesis for the dominant role of particle-bound deposition by showing that the deposition velocities to the plants calculated with the bulk deposition were similar for all PCDD/Fs, but they chose to present their data on a TEQ basis only.

The discussion above serves to illustrate that the relative contributions of the different forms of deposition will be a function of location. Particle-bound deposition can be expected to be more important where the atmospheric particle load, particularly for the larger particle sizes, is higher. Similarly, the influence of wet deposition will depend on the frequency and intensity of precipitation. Plant characteristics will also be important, with particle-bound deposition playing a larger role in those plants which are effective at trapping and retaining particles. Nevertheless, the evidence to date indicates that dry gaseous deposition is the

[35] K. Welsch-Pausch, G. Umlauf and M. S. McLachlan, *Organohalogen Compd.*, 1993, **12**, 99.

[36] G. Rippen and H. Wesp, *Organohalogen Compd.*, 1993, **12**, 111.

dominant pathway of the lower chlorinated PCDD/Fs to grass under rural conditions in Central Europe. Since grass is the main vector of PCDD/Fs into the human food supply and the lower chlorinated PCDD/Fs are of primary concern to human health, dry gaseous deposition would appear to be the most important process.

Dry Gaseous Deposition

Dry gaseous deposition is a partitioning process between the plant and the gas phase. A simple mathematical model with a plant compartment and an air compartment can be written to describe this process.[37] While there is evidence that more than one plant compartment is necessary to describe the uptake behaviour in some species,[38] the one compartment assumption would appear to be reasonable for ryegrass.[39] The structure of the model is illustrated in Figure 3. The defining equations are as follows:

$$\frac{dc_P}{dt} = k_{AP}A(c_{Ag} - c_P/K_{PA})/V \tag{1}$$

where c_P is the concentration in grass (mol m^{-3}), t is time (h), k_{AP} is the air/plant mass transfer coefficient referenced to the gas phase (m h^{-1}), A is the surface area of the plant (m^2), c_{Ag} is the gaseous concentration in air (mol m^{-3}), V is the plant volume (m^3), and K_{PA} is the plant/air partitioning coefficient on a volume/volume basis. k_{AP} is defined by

$$\frac{1}{k_{AP}} = \frac{1}{k_A} + \frac{1}{k_P K_{PA}} \tag{2}$$

where k_A is the mass transfer coefficient from the free atmosphere to the surface of the plant, and k_P is the mass transfer coefficient from the plant surface to the contaminant reservoir in the plant. The inverse of the mass transfer coefficient is a resistance, and in the following these two terms will be referred to as the air-side and plant-side resistances.

For a given plant and a given gaseous concentration there are three unknowns in this model: the air-side resistance, the plant-side resistance and the plant/air partition coefficient. There is little information in the literature on the magnitudes of these parameters for agricultural plants. The one exception is ryegrass, an important pasture grass for which partition coefficients for a number of semi-volatile organic compounds (SOCs) have been measured.[39] A linear relationship was proposed to describe the plant/air partition coefficient as a function of the octanol/air partition coefficient K_{OA} (see Figure 4):

$$K_{PA} = 0.01 \, K_{OA} \tag{3}$$

[37] M. S. McLachlan, K. Welsch-Pausch and J. Tolls, *Environ. Sci. Technol.*, 1995, **29**, 1989.
[38] H. Hauk, G. Umlauf and M. S. McLachlan, *Environ. Sci. Technol.*, 1994, **28**, 2372.
[39] J. Tolls and M. S. McLachlan, *Environ. Sci. Technol.*, 1994, **28**, 159.

Figure 3 Schematic of a simple model of air/plant partitioning

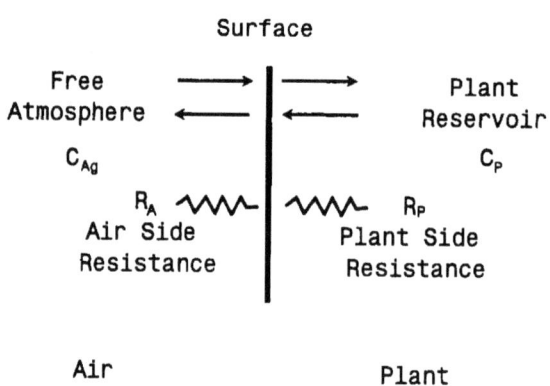

Figure 4 Plot of the measured plant/air partitioning coefficient of several SOCs in ryegrass as a function of the octanol/air partition coefficient (Modified from ref. 39)

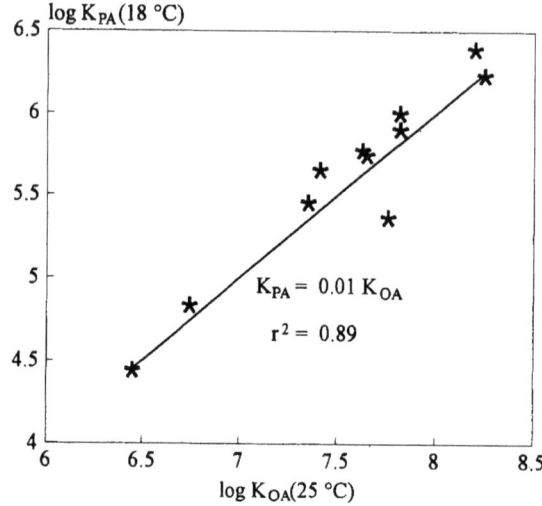

While this equation was obtained for compounds that are somewhat more volatile than the PCDD/Fs, it is consistent with theoretical models of plant uptake of SOCs[40-43] and it would seem reasonable to extrapolate it to the PCDD/Fs.

The same authors also measured the plant-side resistance for the same SOCs in ryegrass. They found no apparent relationship between the magnitude of the plant-side resistance and the molecular properties.[39] This is in agreement with studies of the plant uptake of SOCs in aqueous systems, where it was found that the plant-side resistance was relatively independent of the hydrophobicity for compounds of similar molecular volume.[44] The average value for k_p in ryegrass

[40] K.-W. Schramm, A. Reischl and O. Hutzinger, *Chemosphere*, 1987, **16**, 2653.
[41] M. Riederer, *Environ. Sci. Technol.*, 1990, **24**, 829.
[42] S. Trapp, M. Matthies, I. Scheunert and E. M. Topp, *Environ. Sci. Technol.*, 1990, **24**, 1246.
[43] S. Paterson, D. Mackay and A. Gladman, *Chemosphere*, 1991, **23**, 539.
[44] H. Bauer and J. Schönherr, *Pestic. Sci.*, 1992, **35**, 1.

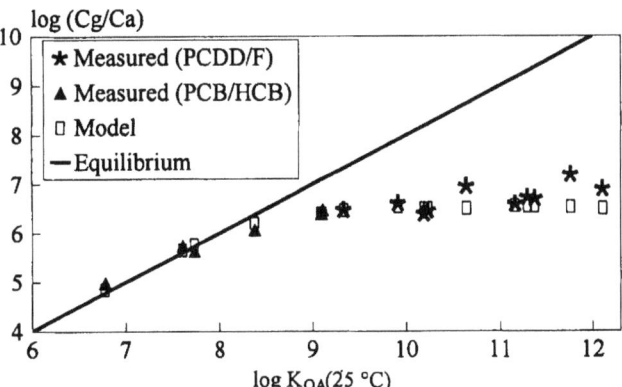

Figure 5 Comparison of measured and modelled plant/air concentration quotients for PCBs and PCDD/Fs as a function of the octanol/air partition coefficient (Modified from ref. 37)

was 2.8×10^{-6} m h^{-1} on a whole plant basis.[39] Since the SOCs studied in this work have molecular volumes similar to the PCDD/Fs, this should be a reasonable estimate for the PCDD/Fs as well.

Having established estimates for two of the three unknowns, the simple model of dry gaseous deposition shown in Figure 3 was applied to a dataset consisting of PCB and PCDD/F concentrations from a ryegrass culture that had been grown outside in flower boxes and from air samples which had been collected during the growing period.[37] The third model parameter, the air-side resistance, was fitted, and the modelled and measured grass concentrations were compared. The agreement was excellent, with maximum deviations between the measured and modelled values of 30% for almost all compounds. Only the Cl_7–Cl_8DD/F displayed higher deviations, with the model underpredicting the measured concentrations. This was attributed to particle-bound deposition which was not accounted for in the model.

In Figure 5 the log of the quotient of the grass and gaseous concentrations for this experiment is plotted against log K_{OA}. In the left part of the plot the grass/gas phase concentration quotients increase linearly with K_{OA}, indicating partitioning equilibrium in accordance with equation (3). However, at higher K_{OA} values the behaviour changes, and the grass/gas phase concentration quotients reach a more or less constant value that does not change with increasing K_{OA}. The grass/gas phase concentration quotients lie below the equilibrium partitioning coefficients given by the extension of the curve on the left of the figure. This is indicative of a non-equilibrium state, and all of the PCDD/Fs lie in this portion of the plot. This is an important result, since it indicates that the resistance and not the partition coefficient will determine PCDD/F accumulation in vegetation.

Of the two resistances that limited uptake, the model indicates that the air-side resistance was dominant.[37] Expressed in simple terms, for compounds with partition coefficients in the order of 10^7 or more, the plant does not 'see' enough air in its lifetime to allow it to achieve equilibrium. Given this and the fact that the PCDD/Fs were orders of magnitude away from equilibrium, equation (1) can be simplified to

$$\frac{dc_P}{dt} = k_A A c_{Ag}/V \tag{4}$$

or

$$c_P = \frac{1}{V} \int k_A A c_{Ag}\, dt \tag{5}$$

This states that the concentration of the lower chlorinated PCDD/Fs in vegetation is the time integral of the product of the air-side mass transfer coefficient (the inverse of the air-side resistance), the specific surface area of the vegetation and the gaseous air concentration.

Can this equation be extended to other plant species? While the non-equilibrium accumulation behaviour has to date only been demonstrated for this ryegrass culture, it is likely also true for other agricultural plants since the PCDD/Fs in the ryegrass were orders of magnitude away from equilibrium although the deposition velocities were higher than under natural conditions due to the elevated exposure of the flower boxes. It is improbable that the plant-side resistance could dominate the air-side resistance due to the very high K_{PA} values for the PCDD/Fs (see equations 2 and 3). It may be that particle-bound deposition makes a significant contribution to the levels of lower chlorinated PCDD/Fs for plants that have much better particle trapping properties than ryegrass, or under atmospheric conditions with much higher particle-bound deposition velocities than in Bayreuth where this study was conducted. However, it is felt that equation (5) will be valid for the Cl_4–Cl_5DD/Fs for most agricultural crops under most conditions.

The one remaining unknown in equation (5) is the air-side mass transfer coefficient, which is often referred to as a deposition velocity. It describes the limitations on the transport of gaseous PCDD/Fs from the free atmosphere to the surface of the plant. It can be viewed as having two components: a turbulent resistance describing the transport from the free atmosphere to the laminar boundary layer surrounding the leaf, and a laminar resistance describing the transport through the laminar boundary layer to the surface of the cuticle. As a whole, the deposition velocity is a function of many factors, including the wind speed, temperature, solar radiation, atmospheric stability, canopy structure and leaf surface structure. For a given situation it is, however, very similar for all PCDD/Fs, since it depends only on the diffusivity of the molecule in air. Diffusivity in air is approximately proportional to the square root of the molecular weight, and since this varies by only 50% between Cl_4DF and Cl_8DD, the diffusivity can be expected to vary by less than 25%.

Air-side deposition velocities have been studied extensively for the uptake of inorganic gases in plants. However, the deposition of PCDD/Fs introduces a new dimension to this subject, since these compounds are deposited primarily to the surface of the cuticle,[45] not to the stomatal openings as in the case of most inorganic trace gases. Hence the receptor surface is fundamentally different for

[45] A. Reischl, H. Thoma, M. Reissinger and O. Hutzinger, *Naturwissenschaften*, 1987, **74**, 88.

the PCDD/Fs, and the deposition velocities for inorganic gases in the literature are of limited use for predicting k_A in equation (5).

There is little information in the literature on deposition velocities of PCDD/Fs or other SOCs to plants. The value measured for the flower box culture experiment described above cannot be extrapolated to environmental conditions. The only values the author is aware of were reported for hay and corn silage grown near to Bayreuth in 1989.[46] The data were presented as air scavenging ratios, the amount of air that 1 g of plant material scavenged during its lifetime. When the values of 9 m^3 g^{-1} dw and 4.5 m^3 g^{-1} dw for grass and corn are converted to average deposition velocities, values of 0.0006 m s^{-1} and 0.00018 m s^{-1} are obtained. Note that these are deposition velocities to the canopy, not to the leaf surface, and in this case A in equation (5) has to be defined as canopy area occupied by the plant and not its surface area. These values represent first estimates only, and more research is needed to measure deposition velocities of PCDD/Fs to crops under typical agricultural conditions.

Other interpretations of dry gaseous deposition of PCDD/Fs to plants have been presented in the literature. In one paper, different plants were placed in a chamber and exposed to 2,3,7,8-Cl$_4$DD.[47] The uptake constants were measured and it was found that they correlated well with the specific surface area of the plant. This is in agreement with equation (5) (assuming that the specific surface area A/V has no effect on the deposition velocity k_A under the chamber conditions), but the author attributed his results to a plant-side instead of an air-side resistance. In an earlier paper this author had used uptake rates and clearance rates measured under conditions of different turbulence to calculate plant/gas phase partitioning coefficients for 2,3,7,8-Cl$_4$DD.[48] This approach and the models that have been developed from it[49,50] should be reconsidered in light of the fact that the uptake and clearance rates are dominated by the air-side resistance (and hence by the turbulence).

Particle-bound Deposition

As described above, wet and dry particle-bound deposition are likely important for the accumulation of the higher chlorinated PCDD/Fs in aerial vegetation. The accumulation of particle-bound PCDD/Fs in plants is a function of a myriad of factors. The deposition rate itself is influenced by the particle size spectrum in the atmosphere and the distribution of the PCDD/Fs on the different particle size fractions, and further by the atmospheric turbulence, the canopy and plant properties, and the frequency and intensity of precipitation. The retention of the contaminants on the plant surface depends on the degree to which the particles are permanently retained on the plant and, for those particles which are not retained, the degree of transfer of PCDD/Fs from the particles to the plant cuticle. This is a very complex system that is not yet well understood. One approach that

[46] M. S. McLachlan, *Organohalogen Compd.*, 1995, **26**, 105.
[47] J. K. McCrady, *Chemosphere*, 1994, **28**, 207.
[48] J. K. McCrady and S. P. Maggard, *Environ. Sci. Technol.*, 1993, **27**, 343.
[49] M. Lorber, D. Cleverly, J. Schaum, L. Phillips, G. Schweer, *et al., Sci. Total Environ.*, 1994, **156**, 39.
[50] M. Lorber, *Organohalogen Compd.*, 1995, **24**, 179.

has been taken to predict deposition of PCDD/Fs is to use particle deposition models and particle erosion models that were developed using radioactive tracers.[32,49] However, the extrapolation of these models to PCDD/Fs can be questioned since these compounds will be associated with particles having different sizes and properties than radionuclide particles. A second approach is to use field data to obtain average long-term net deposition velocities of particle-bound PCDD/Fs. For corn and hay it was reported that the gaseous deposition velocities and the net particle bound deposition velocities were very similar (*i.e.* 0.00016 and 0.0006 m s^{-1} respectively; see above).[46] While these values provide a first estimate, much remains to be learned about particle-bound deposition of PCDD/Fs to plants.

Persistence of PCDD/Fs in Plants

The persistence of PCDD/Fs in plants has not been extensively investigated to date. There is little reason to expect that biodegradation is relevant, given that the PCDD/Fs are found primarily in the non-viable cuticle and that they are very resistant to microbial degradation.

There has been one report of photodegradation of 2,3,7,8-Cl$_4$DD in reed canarygrass that was exposed to sunlight following gaseous contamination.[48] However, in another study, no differences were observed between the PCDD/F concentrations in two pasture grass cultures exposed to background air levels, one in the presence of and one protected from UV radiation, which led the authors to conclude that photodegradation is not relevant under natural conditions.[51] The contradictions between these two studies have yet to be resolved.

Summary

Atmospheric deposition is the dominant source of PCDD/Fs in vegetation. Dry gaseous deposition is the primary pathway of the lower chlorinated PCDD/Fs to agricultural plants, while particle-bound deposition is likely a significant mechanism for the higher chlorinated congeners. Dry gaseous deposition is a diffusive partitioning process, but in the case of the PCDD/Fs a partitioning equilibrium is not approached. The uptake is limited by an air-side resistance, and hence atmospheric turbulence is the primary factor besides the gaseous concentration and time in determining the concentrations of the lower chlorinated congeners in vegetation. Particle-bound deposition is a more complex process. There is some evidence that gaseous and net particle-bound deposition velocities to plants are similar, but much more work is needed to obtain good estimates of these parameters.

4   Transfer from Plants to Livestock and Animal Food

Having addressed the accumulation of PCDD/Fs in plants, this section will discuss the transfer from plants to animals, focusing in particular on cattle. First,

[51] K. Welsch-Pausch and M. S. McLachlan, *Organohalogen Compd.*, 1995, **24**, 509.

the sources of PCDD/Fs to livestock will be reviewed. Thereafter the persistence of different congeners in livestock will be examined. This will be followed by a discussion of plant/animal transfer of PCDD/Fs under steady-state and non-steady-state conditions. Finally, a simple model of contaminant accumulation in an agricultural food chain will be discussed.

Sources of PCDD/Fs to Livestock

As was mentioned in the introduction, it has been shown that feed is the primary source of PCDD/Fs to cattle. Water and air make negligible contributions to exposure.[52] The same can be expected for other livestock, given the very low vapour pressure and hydrophobic nature of this class of compounds.

Based on the discussion of PCDD/F uptake in plants, it can be expected that plant components with a large exposed surface area per unit mass will have higher concentrations of PCDD/Fs and be more important vectors to livestock than feeds which develop protected from the atmosphere. This was investigated for the feed ration of cows in Bayreuth and found to be true, with grass products making the dominant contribution to PCDD/F uptake followed by corn silage, while concentrate consisting primarily of grain was of little importance.[6] The more pronounced accumulation in leafy plants is reflected in the PCDD/F concentrations found in different types of meat. Ruminants which consume large quantities of grass have higher PCDD/F levels than pigs, for instance, which consume little if any leafy material.[53]

Another pathway of PCDD/Fs to livestock that has attracted considerable interest is the direct ingestion of soil by grazing animals. In a recent review it was emphasized that there has been a tendency to use unrealistically high values for soil uptake.[54] While soil ingestion rates between <1 and 18% have been measured in sheep and cows, the higher values were associated with winter grazing and no supplemental feed. It is stated that soil ingestion rates of 1–2% of dry matter intake are unlikely to be exceeded when modern agricultural practices involving supplementary feeding are employed, at least in areas where climatic conditions prevent year-round grazing. In another paper the average soil ingestion rate of grazing milk cattle in Holland was estimated to be $225\,\mathrm{g\,d^{-1}}$, about 1.2% of the total dry feed intake.[27] It is interesting to note that this is less than the same author's estimate of soil intake in winter from grass silage, suggesting that the cow is more selective than the harvesting machine.

This soil ingestion rate was used together with an estimate of the background level of feed contamination in northwestern Germany in 1990 to compare the relative contributions of soil and feed to uptake of PCDD/Fs by cattle. It was found that the soil concentration would have to be $47\,\mathrm{ng\,TEQ\,kg^{-1}}$ for the uptake from soil to equal the uptake from feed.[34] However, the concentrations of PCDD/Fs in air are decreasing in Germany.[55] This is being accompanied by a decrease in feed concentrations, which is reflected in the pronounced decrease in

[52] M. S. McLachlan, H. Thoma, M. Reissinger and O. Hutzinger, *Chemosphere*, 1990, **20**, 1013.
[53] O. Päpke and P. Fürst, *Organohalogen Compd.*, 1995, **22**, 143.
[54] G. F. Fries, *Sci. Total Environ.*, 1996, **185**. 93.
[55] E. Hiester, P. Bruckmann, R. Böhm, P. Eynck, A. Gerlach, *et al.*, *Organohalogen Compd.*, 1995, **24**, 147.

milk concentrations.[56] If this trend continues the relative importance of soil ingestion and soil contamination of feed as a source of PCDD/Fs to the food chain will increase. The decrease in feed contamination will have to be close to an order of magnitude though, before typical agricultural soils (1–5 ng TEQ kg^{-1}) begin to contribute significantly to contamination of dairy products and beef.

Soil uptake can have a more significant impact on PCDD/F concentrations in eggs from free-ranging hens.[57,58] In a survey of eggs conducted in northwestern Germany, eggs from free-ranging hens had considerably higher concentrations than eggs from hens housed in elevated cages.[59] This can be attributed to the high soil ingestion of foraging hens. While this pathway is of little relevance for the general food supply in industrialized countries, it can be important for small sub-populations.

Persistence of PCDD/Fs in Livestock

In general, only 2,3,7,8-substituted PCDD/Fs are found in livestock. The results of a mass balance of a lactating cow indicated that the non-2,3,7,8-substituted PCDD/Fs are absorbed from the feed in the digestive tract and then metabolized.[52] This study also showed that several of the 2,3,7,8-substituted dibenzofurans are largely metabolized in cattle, namely 2,3,7,8-Cl_4DF and 1,2,3,7,8-Cl_5DF. It is suspected that 1,2,3,7,8,9-Cl_6DF is also metabolized, but the low concentrations of this congener have frustrated the resolution of this question. The metabolism of 2,3,7,8-Cl_4DF and 1,2,3,7,8-Cl_5DF is also apparent from feeding studies in which the half-life of these congeners in the cow were orders of magnitude lower than the half-lives of other 2,3,7,8-substituted PCDD/Fs.[60,61] A similar behaviour has been reported for hens.[62]

Steady-state Transfer in Livestock

A steady state exists when the uptake, storage and excretion of a substance does not change with time. Given that the concentrations of PCDD/Fs in feed are relatively constant, milk cows and egg laying hens are in an approximate PCDD/F steady state. The one exception is the cow that has calved for the first time. In this case the PCDD/Fs that it has accumulated as a heifer are initially

[56] O. Fürst and K. Wilmers, *Organohalogen Compd.*, 1995, **26**, 101.

[57] R. R. Chang, D. G. Hayward, L. R. Goldman, M. Harnly, J. J. Flattery, *et al.*, *Chemosphere*, 1989, **19**, 481.

[58] R. D. Stephens, M. Harnly, D. G. Hayward, R. R. Chang and J. J. Flattery, *Chemosphere*, 1990, **20**, 1091.

[59] P. Fürst, C. Fürst and K. Wilmers, *Organohalogen Compd.*, 1993, **13**, 31.

[60] M. Olling, H. J. G. M. Derks, P. L. M. Berende, A. K. D. Liem and A. P. J. M. de Jong, *Chemosphere*, 1991, **23**, 1377.

[61] W. Heeschen, A. Blüthgen and U. Ruoff, *Untersuchungen zum Übergang ausgewählter polychlorierter Dibenzo-para-dioxine und -Furane nach oraler Supplementierung in die Milch laktierender Kühe*, Bundesanstalt für Milchforschung, Kiel, 1994.

[62] R. D. Stephens, M. X. Petreas, D. G. Hayward, R. R. Chang, L. R. Goldman, *et al.*, poster presented at the 11th International Symposium on Chlorinated Dioxins and Related Compounds, Research Triangle Park, 23–27 September, 1991.

excreted through the milk, and a steady state is only approached sometime during the first lactation period.[63]

At steady state the excretion of a persistent compound equals the uptake. In cows and hens, feed/soil is the only relevant source of PCDD/F uptake, and there are two forms of excretion: faeces and milk/eggs. If, for a given PCDD/F uptake, the flux in the faeces can be determined, then the net transfer of PCDD/Fs from the feed to the food product (often referred to as carryover) can be easily calculated. Or, in other words, at steady state all of the persistent PCDD/Fs that are absorbed by the animal are transferred to the milk/eggs. Hence, we need to understand absorption in order to understand carryover.

A mass balance of a lactating cow was conducted to measure absorption of PCDD/Fs.[52,64] It was found that the absorption of the PCDD/Fs decreased with increasing degree of chlorination. This decrease was attributed to the increasing hydrophobicity of the PCDD/Fs in a model of contaminant transfer in the cow.[65]

While this was the only study that directly investigated absorption, there have been a number of reports on feed to milk transfer. These are summarized in Table 2 as carryover rates, defined as the fraction of the ingested contaminant that is excreted in the milk at steady state. The data in the first four papers were obtained from feeding studies in which cows were fed PCDD/Fs, usually in capsule form, and the steady-state excretion rate was extrapolated from the uptake and clearance kinetics.[32,60,61,66] The fifth column gives the results of the mass balance study,[64] and the data in the sixth column were obtained from a field study.[67] There is generally good agreement between the studies, the carryover ranging from about 0.35 for the lower chlorinated persistent congeners down to several percent at most for Cl_8DD/F.

There is very little information on how the form in which the PCDD/Fs are present in the feed influences the absorption and hence carryover. Table 2 shows good agreement between the feeding studies in which the PCDD/Fs were artificially added to the feed and the mass balance study where the only source of PCDD/Fs was from naturally contaminated feed from a rural area. However, the carryover rate and hence the absorption were lower in the field study, which was conducted near a known source of PCDD/Fs. It was suggested that the lower carryover in the field study was due to a lower absorption of PCDD/Fs from flyash that had been deposited on the feed following emission from a nearby incinerator. This hypothesis was supported with the results of a laboratory study in which cows were fed PCDD/Fs on fly ash. The carryover in this case was more than an order of magnitude lower than in the field study.[67] This shows that the source of the PCDD/Fs can influence absorption in the cow. It remains to be seen whether the carryover of PCDD/Fs in soil, sewage sludge or other specific

[63] M. S. McLachlan, in *Kriterien zur Beurteilung organischer Bodenkontaminationen: Dioxine (PCDD/F) und Phthalate*, ed. G. Kreysa and J. Wiesner, Dechema, Frankfurt a. M., 1995, p. 309.

[64] M. S. McLachlan, *Das Verhalten hydrophober chlororganischer Verbindungen in laktierenden Rindern*, Doctoral Thesis, University of Bayreuth, 1992.

[65] M. S. McLachlan, *Environ. Sci. Technol.*, 1994, **28**, 2407.

[66] D. Firestone, M. Clower, Jr., A. P. Borsetti, R. H. Teske and P. E. Long, *J. Agric. Food Chem.*, 1979, **27**, 1171.

[67] W. Slob, M. Olling, H. J. G. M. Derks and A. P. J. M. de Jong, *Chemosphere*, 1995, **31**, 3827.

Table 2 Carryover rates of 2,3,7,8-substituted PCDD/Fs from feed to milk

Congener	Ref 32	Ref 60	Ref 61	Ref 66	Ref 64	Ref 67
2,3,7,8-Cl$_4$DD	0.40		0.30	0.35–0.60	0.36	0.15
1,2,3,7,8-Cl$_5$DD			0.28	0.10–0.15	0.32	0.10
1,2,3,4,7,8-Cl$_6$DD				0.05–0.10	0.16	0.06
1,2,3,6,7,8-Cl$_6$DD		0.33	0.27	0.10–0.15	0.15	0.06
1,2,3,7,8,9-Cl$_6$DD				0.05–0.10	0.15	0.03
1,2,3,4,6,7,8-Cl$_7$DD		0.03	0.02		0.03	0.006
Cl$_8$DD		0.01			0.04	0.001
2,3,7,8-Cl$_4$DF			0.01	<0.01	0.07	0.008
1,2,3,7,8-Cl$_5$DF				0.01–0.05	0.05	0.005
2,3,4,7,8-Cl$_5$DF			0.36	0.25–0.35	0.33	0.12
1,2,3,4,7,8-Cl$_6$DF			0.18	0.05–0.10	0.15	0.04
1,2,3,6,7,8-Cl$_6$DF					0.15	0.04
2,3,4,6,7,8-Cl$_6$DF					0.14	0.04
1,2,3,4,6,7,8-Cl$_7$DF			0.02	<0.01	0.03	0.004
1,2,3,4,7,8,9-Cl$_7$DF					0.08	0.005
Cl$_8$DF					0.02	

PCDD/F sources differs considerably from the carryover in 'naturally' contaminated feed.

There has been considerably less work on steady-state carryover in hens. However, in one detailed study it was shown that the absorption of PCDD/Fs from a contaminated soil was quite high, ranging from 70–80% for the tetrachlorinated congeners to 10% for the octachlorinated ones.[68] The authors compared their results to feeding studies with rodents and came to the conclusion that absorption from soil was as effective as absorption from easily digestible carriers such as corn oil.

Non-steady-state Transfer in Livestock

Whereas milk cows and egg laying hens are at a near steady state, animals raised for meat production such as beef cattle and pigs are not. These animals do not have the possibility to excrete the PCDD/Fs in a lipid-rich matrix such as milk or eggs. All of the persistent contaminant that is absorbed is stored in the body. Although they could theoretically reach a steady state in which the net absorption of persistent PCDD/Fs was zero, this condition is not approached under modern animal husbandry practices. The life span of the animals is short and the growth rate is high, serving constantly to dilute the contaminant stores in the animal. Indeed, PCDD/F concentrations in beef fat are generally quite similar to those in milk fat,[2,3] although the steady-state concentration in beef cattle has been estimated to be at least an order of magnitude higher than in lactating cows.[65]

Several approaches have been used to predict the transfer of persistent organic contaminants from feed to beef. One involves simply using a bioconcentration or biotransfer factor, a ratio of the concentration in animal fat to the concentration or uptake in the feed.[49] This method has the disadvantage that it does not account for variability in animal husbandry practices. A more sophisticated

---

[68] R. D. Stephens, M. X. Petreas and D. G. Hayward, *Organohalogen Compd.*, 1994, **20**, 55.

approach that provides detailed information is to conduct a model simulation of the accumulation in the animal. For instance, it was shown for pigs that the residue concentration peaks at about 100 days of age and then decreases due to the shift in energy partitioning from protein deposition to fat deposition with increasing age.[69] Recently, a method has been proposed which is simple but still accounts for the most important variables. The total amount of a persistent contaminant ingested by an animal over its lifetime is multiplied by the absorption rate and divided by the weight of lipid in the animal at slaughter to obtain the lipid-based concentration.[65] This approach makes use of the fact that the PCDD/Fs are stored almost exclusively in body lipids.

Another aspect of non-steady-state behaviour that is occasionally of interest is the clearance of a chemical from a contaminated animal. Feeding studies have shown the clearance from non-lactating cattle to be very slow, with half-lives in the order of several hundred days.[70,71] It is not known if the disappearance is due to metabolism or other mechanisms. While the loss from the animal is very slow, the concentrations can decrease as a result of growth dilution.[72]

Clearance from lactating cows has been studied much more intensively. Half-lives in the order of 30–60 days have generally been found.[60,66,73,74] The half-life can be expected to increase with increasing body lipid weight and decreasing lactation rate of the contaminated animal.[65] There is evidence of a rapid initial drop in milk fat concentrations following removal of the contaminant source from the diet, and this has been described using a pharmacokinetic model.[75]

A Simple Model of PCDD/F Accumulation in an Agricultural Food Chain

The knowledge about the uptake of PCDD/Fs by plants was combined with information on the transfer from feed to animal products to assemble a predictive model of PCDD/F accumulation in the most important agricultural food chain for human exposure to these compounds, namely air–fodder–cow–milk.[46] The air/plant scavenging ratios of $9 m^3 g^{-1}$ dw for grass and $4.5 m^3 g^{-1}$ dw for corn silage were multiplied by typical grass and corn silage rations for dairy cows in Bavaria and added. The resulting number, 100 000, gives the equivalent volume of air cleaned by the feed that a cow eats daily (*i.e.* the cow 'eats' the PCDD/Fs in $100 000 m^3$ of air). Multiplying this by the air concentration c_A (gaseous + particle bound) yields the daily uptake of the PCDD/F congener by the cow. This is then multiplied by the carryover rates from the mass balance study (see Table 2) and

[69] G. F. Fries, *Organohalogen Compd.*, 1994, **20**, 51.
[70] D. J. Jensen, R. A. Hummel, N. H. Mahle, C. W. Kocher and H. S. Higgins, *J. Agric. Food Chem.*, 1981, **29**, 265.
[71] M. Olling, P. L. M. Berende, H. J. G. M. Derks, A. K. D. Liem, H. Everts, *et al., De toxikokinetiek van PCDD's en PCDF's in niet lacterende koeien (vetweiders)*, Rijksinstituut voor Volksgezondheid en Milieuhygiene, Rapport 328904003, Bilthoven, 1991.
[72] J. R. Startin, C. Wright, M. Kelly and N. Harrison, *Organohalogen Compd.*, 1994, **21**, 347.
[73] D. J. Jensen and R. A. Hummel, *Bull. Environ. Contam. Toxicol.*, 1982, **29**, 440.
[74] A. Riss and H. Hagenmaier, *VDI Ber.*, 1991, **901**, 863.
[75] H. J. G. M. Derks, P. L. M. Berende, M. Olling, H. Everts, A. K. D. Liem, *et al., Chemosphere*, 1994, **28**, 711.

divided by an average daily milk fat lactation rate to give the milk fat concentration c_M:

$$c_M = c_A \times 100\,000 \times \text{carryover/lactation (g milk fat d}^{-1}) \qquad (6)$$

This simple model was tested using data from air samples collected in Bayreuth during the summer of 1989. The predicted milk concentrations were compared with the average concentrations obtained in a survey of over 100 milk samples collected in Bavaria in 1989 and 1990. There was excellent agreement between the predicted and measured milk concentrations for almost all congeners (see Figure 6). This suggests that there is now a good basic understanding of the accumulation of these compounds in agricultural food chains.

5 Transfer from Food to Humans

The primary sources of human exposure to PCDD/Fs were outlined in the introduction. For North America and Western Europe, uptake of these compounds is about equally divided between dairy, meat and fish products (see Table 1). The levels in the diet are decreasing in some countries. The most recent estimates for Germany suggest an intake of 55–70 pg TEQ d^{-1} for adults.[76]

Of crucial importance for the potential risk associated with this contamination of the diet is the question: how much of the ingested PCDD/Fs is actually absorbed and retained in the body? There is, however, remarkably little information on this subject. There is one report of absorption in an adult man where 87% of a single dose of radioactively labelled 2,3,7,8-Cl$_4$DD dissolved in corn oil was taken up.[77] In another study, in which only faeces samples from two individuals on an unmodified diet were analysed, it was found that the excretion rate was very high compared to the theoretical uptake rate,[78] which would suggest that absorption is low. No work beyond these two somewhat contradictory studies was found.

Of particular interest is the question of whether or not an absorptive steady state is approached. In this case the net absorption of a compound from the normal diet would be near zero. Human tissue levels might then be determined primarily by highly contaminated foods which are seldom consumed but nevertheless result in a significant absorption of the compound. This would have dramatic consequences for the way risk analysis is done and the identification of means to reduce the risk associated with PCDD/Fs.[79] There is an urgent need for more research in this area.

It is known that the PCDD/Fs are very persistent in humans. Half-lives of most 2,3,7,8-substituted congeners have been measured in occupationally exposed cohorts and found to lie in the range of 5–15 years.[80] This results in accumulation in humans, and the concentrations in human fat are at least an order of magnitude higher than in cows' milk fat or beef fat.[79] As in livestock, the

[76] M. Grün, O. Päpke, M. Weibrodt, A. Lis, M. Ball, *et al., Organohalogen Compd.*, 1995, **26**, 151.

[77] H. Poiger and C. Schlatter, *Chemosphere*, 1986, **15**, 1489.

[78] R. Andersson and C. Rappe, *Organohalogen Compd.*, 1992, **9**, 195.

[79] M. S. McLachlan, *Environ. Sci. Technol.*, 1996, **30**, 252.

[80] D. Flesch-Janys, P. Gurn, D. Jung, J. Koonietzko, A. Manz, *et al., Organohalogen Compd.*, 1994, **21**, 93.

Figure 6 Comparison of the predicted and measured concentrations of the 2,3,7,8-substituted PCDD/F congeners in milk (Taken from ref. 46)

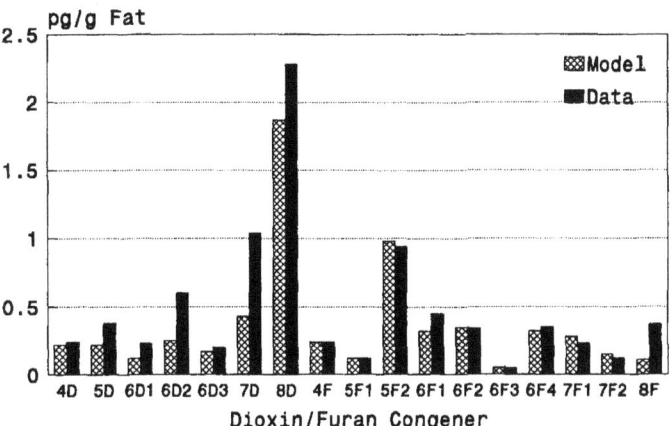

concentrations of 2,3,7,8-Cl$_4$DF and 1,2,3,7,8-Cl$_5$DF are very low in humans, suggesting that these congeners are metabolized.

6 Transfer from Human Milk to Infants

Nursing infants are exposed to unusually high levels of PCDD/Fs. As their food source is human milk, they are effectively one link higher in the food chain than the rest of the population. At the end of the 1980s it was estimated that nursing infants in developed countries ingest about 85 pg of 2,3,7,8-Cl$_4$DD toxicity equivalents (TEQs) per kg body weight and day.[81] While there is evidence that the mothers' milk concentrations have decreased in the meantime,[53,82] infant exposure is still far in excess of the recommended acceptable daily intakes (ADIs), which range from 0.0064 to 10 pg TEQ kg^{-1} BW day, depending on the jurisdiction.[83]

In contrast to adults, in nursing infants the absorption of PCDD/Fs has been well studied. All investigators have come to the same conclusion, namely that the absorption of PCDD/Fs is nearly complete.[84–88]. The high exposure of infants continues to be one of the biggest concerns in the dioxin issue.

7 Concluding Remarks

To review, PCDD/Fs enter the food chain primarily from the atmosphere. Diffusion of gaseous PCDD/Fs to the lipophilic cuticle is the primary mechanism of uptake into plants, although the deposition of particle-bound chemical likely

[81] WHO, *Levels of PCBs, PCDDs, and PCDFs in Breast Milk*, WHO Regional Office for Europe, FADL, Copenhagen, 1989.

[82] P. Fürst, C. Fürst and K. Wilmers, *Chemosphere*, 1992, **25**, 1029.

[83] E. W. Bretthauer, H. W. Kraus and A. di Domenico, *Dioxin Perspectives. A Pilot Study on International Information Exchange on Dioxins and Related Compounds*, Plenum Press, New York, 1991.

[84] B. Jödicke, M. Ende, H. Helge and D. Neubert, *Chemosphere*, 1992, **25**, 1061.

[85] W. Körner, N. Dawidowsky and H. Hagenmaier, *Chemosphere*, 1993, **27**, 157.

[86] H. J. Pluim, J. Wever, J. G. Koppe, J. W. vd Slikke and K. Olie, *Chemosphere*, 1993, **26**, 1947.

[87] M. S. McLachlan, *Toxicol. Appl. Pharmacol.*, 1993, **123**, 68.

[88] P. Dahl, G. Lindström, K. Wiberg and C. Rappe, *Chemosphere*, 1995, **30**, 2297.

plays a significant role for the higher chlorinated congeners. Although dry gaseous deposition is a partitioning process, a partitioning equilibrium is not approached. The uptake is limited by the rate of transport of chemical to the plant surface, and thus the accumulation of PCDD/Fs in plants is primarily a function of the gaseous concentration, the turbulence in the atmosphere and the length of exposure. PCDD/Fs in plants are ingested by livestock. The absorption of the different congeners is variable, decreasing from relatively high values of 30–80% for the Cl_4–Cl_5DD/F to at most several percent for the Cl_8DD/Fs. Once in the animal, most of the 2,3,7,8-substituted congeners are persistent and either accumulate in fatty tissue or, in the case of cows and hens, are excreted in milk and eggs. From there they are ingested by humans. Little is known about the degree of absorption in adults, but PCDD/Fs accumulate to much higher levels in humans than in livestock. This contaminant is transferred to nursing infants through the mothers' milk, and here it has been shown that the ingested PCDD/Fs are virtually completely absorbed. Consequently, nursing infants are subject to the highest daily dose on a body weight basis.

The biological uptake and transfer of PCDD/Fs in agricultural food chains has also been examined from a bioaccumulation perspective using fugacities.[79] In aquatic food chains, persistent lipophilic organic contaminants are often biomagnified, which means that their fugacity or lipid-based concentration increases as the contaminant moves up the food chain. However, in agricultural food chains there is a biodilution of PCDD/Fs, with the fugacity decreasing by up to three orders of magnitude between air and cows' milk. This can be attributed to the kinetically limited uptake in plants, which prevents the achievement of an air/plant partitioning equilibrium, and to the reduced absorption of the more lipophilic congeners in livestock. This biodilution phenomenon is fortuitous, since if food products were in equilibrium with the air then human exposure to PCDD/Fs would be much higher.

It is important to note that the degree of accumulation of PCDD/Fs in agricultural food chains differs from congener to congener. This is due primarily to the behaviour in livestock, where many congeners are metabolized and others are preferentially transferred to food products owing to more efficient absorption from feed. Consequently, two PCDD/F sources with similar toxicity equivalents might pose very different risks to the food supply and human exposure. For instance, a source contributing primarily 1,2,3,7,8-Cl_5DF would be of much less concern than a source contributing 2,3,4,7,8-Cl_5DF, since most of the former compound would be metabolized in livestock. Hence it is not possible to compare the risk from different sources or contaminated environmental matrices on the basis of the TEQs. This must be done on a congener by congener basis or using exposure toxicity equivalents, a modification of the TEQ system that accounts for the differences in the uptake and transfer behaviour between the PCDD/F congeners.[89]

8 Acknowledgements

The author thanks Kerstin Welsch-Pausch for preparing Figure 1.

[89] M. S. McLachlan, *Chemosphere*, 1993, **27**, 483.

Atmospheric Chemistry of PCBs, PCDDs and PCDFs

ROGER ATKINSON

1 Introduction

Polychlorinated biphenyls (PCBs), polychlorinated dibenzo-*p*-dioxins (PCDDs) and polychlorinated dibenzofurans (PCDFs) (see Figure 1) are emitted into the atmosphere from the combustion of chlorine-containing materials.[1-13] Additionally, PCBs, which were used as lubricants and dielectric fluids,[13] are volatilized into the atmosphere from soils and bodies of water (for example, from the Great Lakes) after their disposal or inadvertent release into, and cycling through, the environment.[1,14-18] Atmospheric concentrations of PCBs, PCDDs and PCDFs have been measured at numerous locations,[2,15,16,19-62] and these organochlorine

[1] T. J. Murphy, L. J. Formanski, B. Brownawell and J. A. Meyer, *Environ. Sci. Technol.*, 1985, **19**, 942.
[2] K. Ballschmiter, H. Buchert, R. Niemczyk, A. Munder and M. Swerev, *Chemosphere*, 1986, **15**, 901.
[3] H. Y. Tong and F. W. Karasek, *Chemosphere*, 1986, **15**, 1219.
[4] H. Hagenmaier, H. Brunner, R. Haag and A. Berchtold, *Chemosphere*, 1986, **15**, 1421.
[5] K. Scheidl, F. Wurst and R.-P. Kuna, *Chemosphere*, 1986, **15**, 2089.
[6] R. E. Clement, H. M. Tosine, J. Osborne, V. Ozvacic, G. Wong and S. Thorndyke, *Chemosphere*, 1987, **16**, 1895.
[7] A. Yasuhura, H. Ito and M. Morita, *Environ. Sci. Technol.*, 1987, **21**, 971.
[8] J. R. Visalli, *J. Air Pollut. Control Assoc.*, 1987, **37**, 1451.
[9] D. Pitea, M. Lasagni, L. Bonati, G. Moro, R. Todeschini and G. Chiesa, *Chemosphere*, 1989, **18**, 1465.
[10] S. Marklund, R. Andersson, M. Tysklind, C. Rappe, K.-E. Egebäck, E. Björkman and E. Grigoriadis, *Chemosphere*, 1990, **20**, 553.
[11] M. Oehme, S. Larssen and E. M. Brevik, *Chemosphere*, 1991, **23**, 1699.
[12] R. Bacher, M. Swerev and K. Ballschmiter, *Environ. Sci. Technol.*, 1992, **26**, 1649.
[13] V. Lang, *J. Chromatogr.*, 1992, **595**, 1.
[14] J. B. Manchester-Neesvig and A. W. Andren, *Environ. Sci. Technol.*, 1989, **23**, 1138.
[15] J. E. Baker and S. J. Eisenreich, *Environ. Sci. Technol.*, 1990, **24**, 342.
[16] D. R. Achman, K. C. Hornbuckle and S. J. Eisenreich, *Environ. Sci. Technol.*, 1993, **27**, 75.
[17] K. C. Hornbuckle, D. R. Achman and S. J. Eisenreich, *Environ. Sci. Technol.*, 1993, **27**, 87.
[18] J. D. Jeremiason, K. C. Hornbuckle and S. J. Eisenreich, *Environ. Sci. Technol.*, 1994, **28**, 903.
[19] E. Atlas and C. S. Giam, *Science*, 1981, **211**, 163.
[20] H. Buchert and K. Ballschmiter, *Chemosphere*, 1986, **15**, 1923.
[21] J. M. Czuczwa and R. A. Hites, *Environ. Sci. Technol.*, 1986, **20**, 195.
[22] B. D. Eitzer and R. A. Hites, *Int. J. Environ. Anal. Chem.*, 1986, **27**, 215.
[23] T. F. Bidleman, W. N. Billings and W. T. Foreman, *Environ. Sci. Technol.*, 1986, **20**, 1038.

compounds have been observed in the atmosphere in remote areas of the world,[19,32,36,40,48-50,56] including over the Pacific Ocean[19,56] and in the Arctic.[32,36,40,48,49] To date, mainly for health risk assessment, the published ambient air measurements of the PCDDs and PCDFs have dealt almost exclusively with the tetrachloro- to the octachlorodibenzo-*p*-dioxins, and dibenzofurans. The observations of these organochlorine compounds in lakes and ocean waters remote from direct sources (for example, those in Arctic regions such as northern Canada and Alaska) confirm the occurrence of long-range

[24] B. J. Fairless, D. I. Bates, J. Hudson, R. D. Kleopfer, T. T. Holloway, D. A. Morey and T. Babb, *Environ. Sci. Technol.*, 1987, **21**, 550.

[25] C. Rappe and L.-O. Kjeller, *Chemosphere*, 1987, **16**, 1775.

[26] T. F. Bidleman, U. Wideqvist, B. Jansson and R. Söderlund, *Atmos. Environ.*, 1987, **21**, 641.

[27] R. Wittlinger and K. Ballschmiter, *Chemosphere*, 1987, **16**, 2497.

[28] C. Rappe, L.-O. Kjeller, P. Bruckmann and K.-H. Hackhe, *Chemosphere*, 1988, **17**, 3.

[29] W. Christmann, K. D. Klöppel, H. Partscht and W. Rotard, *Chemosphere*, 1989, **19**, 521.

[30] J. C. Duinker and F. Bouchertall, *Environ. Sci. Technol.*, 1989, **23**, 57.

[31] T. O. Tiernan, D. J. Wagel, G. F. Vanness, J. H. Garrett, J. G. Solch and L. A. Harden, *Chemosphere*, 1989, **19**, 541.

[32] G. W. Patton, D. A. Hinckley, M. D. Walla and T. F. Bidleman, *Tellus*, 1989, **41B**, 243.

[33] H. Y. Tong, S. Arghestani, M. L. Gross and F. W. Karasek, *Chemosphere*, 1989, **18**, 577.

[34] C. Rappe, S. Marklund, L.-O. Kjeller and A. Lindskog, *Chemosphere*, 1989, **18**, 1283.

[35] S. A. Edgerton, J. M. Czuczwa, J. D. Rench, R. F. Hodanbosi and P. J. Koval, *Chemosphere*, 1989, **18**, 1713.

[36] D. J. Gregor and W. D. Gummer, *Environ. Sci. Technol.*, 1989, **23**, 561.

[37] P. Larsson and L. Okla, *Atmos. Environ.*, 1989, **23**, 1699.

[38] B. D. Eitzer and R. A. Hites, *Environ. Sci. Technol.*, 1989, **23**, 1389.

[39] B. D. Eitzer and R. A. Hites, *Environ. Sci. Technol.*, 1989, **23**, 1396.

[40] T. F. Bidleman, G. W. Patton, M. D. Walla, B. T. Hargrave, W. P. Vass, P. Erickson, B. Fowler, V. Scott and D. J. Gregor, *Arctic*, 1989, **42**, 307.

[41] T. Nakano, M. Tsuji and T. Okuno, *Atmos. Environ., Part A*, 1990, **24**, 1361.

[42] R. Wittlinger and K. Ballschmiter, *Fresenius' J. Anal. Chem.*, 1990, **336**, 193.

[43] G. T. Hunt and B. E. Maisel, *Chemosphere*, 1990, **20**, 1455.

[44] W. T. Foreman and T. F. Bidleman, *Atmos. Environ., Part A*, 1990, **24**, 2405.

[45] T. M. Holsen, K. E. Noll, S.-P. Liu and W.-J. Lee, *Environ. Sci. Technol.*, 1991, **25**, 1075.

[46] K. Ballschmiter and R. Wittlinger, *Environ. Sci. Technol.*, 1991, **25**, 1103.

[47] D. Broman, C. Näf and Y. Zebühr, *Environ. Sci. Technol.*, 1991, **25**, 1841.

[48] M. Oehme, *Sci. Total Environ.*, 1991, **106**, 43.

[49] M. Oehme, *Ambio*, 1991, **20**, 293.

[50] A. H. Knap and K. S. Binkley, *Atmos. Environ., Part A*, 1991, **25**, 1507.

[51] G. T. Hunt and B. E. Maisel, *J. Air Waste Manage. Assoc.*, 1992, **42**, 672.

[52] R. M. Hoff, D. C. G. Muir and N. P. Grift, *Environ. Sci. Technol.*, 1992, **26**, 266.

[53] R. M. Hoff, D. C. G. Muir and N. P. Grift, *Environ. Sci. Technol.*, 1992, **26**, 276.

[54] B. Ngabe and T. F. Bidleman, *Environ. Pollut.*, 1992, **76**, 147.

[55] I. Holoubek, A. Kočan, J. Petrik, J. Chovancova, K. Bliková, P. Kořinek, I. Holoubková, J. Pekárek, F. Kott, A. Pacl, M. Bezačinsky and L. Neubauerová, *Toxicol. Environ. Chem.*, 1992, **36**, 115.

[56] H. Iwata, S. Tanabe, N. Sakai and R. Tatsukawa, *Environ. Sci. Technol.*, 1993, **27**, 1080.

[57] M. Tysklind, I. Fängmark, S. Marklund, A. Lindskog, L. Thaning and C. Rappe, *Environ. Sci. Technol.*, 1993, **27**, 2190.

[58] H. Kaupp, J. Towara and M. S. McLachlan, *Atmos. Environ.*, 1994, **28**, 585.

[59] D. H. White and J. W. Hardy, *Environ. Monitor. Assess.*, 1994, **33**, 247.

[60] S. Y. Panshin and R. A. Hites, *Environ. Sci. Technol.*, 1994, **28**, 2001.

[61] S. Y. Panshin and R. A. Hites, *Environ. Sci. Technol.*, 1994, **28**, 2008.

[62] R. L. Falconer, T. F. Bidleman and W. E. Cotham, *Environ. Sci. Technol.*, 1995, **29**, 1666.

Figure 1 Structures of the parent aromatic and heterocyclic compounds

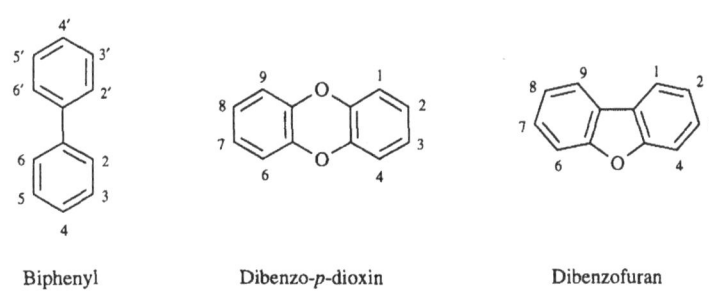

Biphenyl   Dibenzo-*p*-dioxin   Dibenzofuran

transport through the atmosphere.[32,40,63,64] Because these organochlorine compounds are lipophilic,[65] bioaccumulation of PCBs, PCDDs and PCDFs occurs in mammals which feed on fish in such northern lakes and ocean waters, with potentially adverse biological effects.[48,63,64,66-69]

To understand quantitatively the lifetimes and fates of PCBs, PCDDs and PCDFs in the atmosphere, the various potential tropospheric removal processes for these organochlorine compounds need to be assessed and the dominant removal and/or transformation processes identified. In this chapter, the available information concerning the atmospheric removal processes of the PCBs, PCDDs and PCDFs is presented and reviewed.

2 Gas/Particle Partitioning of PCBs, PCDDs and PCDFs in the Atmosphere

As for other chemical compounds,[70-72] in the atmosphere the PCDDs, PCDFs and PCBs are partitioned[14,22,23,30,38,39,41,44,45,47,51,62,73] between the gas and particle phases, with the gas/particle partitioning depending on the liquid-phase (or sub-cooled liquid-phase) vapour pressure at the ambient atmospheric temperature, P_L, the surface area of the particles per unit volume of air (for example, cm^2 of particle per cm^3 of air), θ, and the nature of the particles and of the chemical being adsorbed.[62,70,71] The fraction of the chemical present in the particle phase, ϕ, depends on these parameters through an equation of the form[62,70-72]

$$\phi = c\theta/(c\theta + P_L) \tag{1}$$

where c is a parameter which depends on the chemical being adsorbed and on the

[63] C. R. Macdonald and C. D. Metcalfe, *Can. J. Fish. Aquat. Sci.*, 1991, **48**, 371.
[64] W. L. Lockhart, R. Wagemann, B. Tracey, D. Sutherland and D. J. Thomas, *Sci. Total Environ.*, 1992, **122**, 165.
[65] W. Y. Shiu and D. Mackay, *J. Phys. Chem. Ref. Data*, 1986, **15**, 911.
[66] P. A. Pearce, J. E. Elliott, D. B. Peakall and R. J. Norstrom, *Environ. Pollut.*, 1989, **56**, 217.
[67] R. J. Norstrom, M. Simon and D. C. G. Muir, *Environ. Pollut.*, 1990, **66**, 1.
[68] C. E. Bacon, W. M. Jarman and D. P. Costa, *Chemosphere*, 1992, **24**, 779.
[69] R. J. Norstrom and D. C. G. Muir, *Sci. Total Environ.*, 1994, **154**, 107.
[70] J. F. Pankow, *Atmos. Environ.*, 1987, **21**, 2275.
[71] T. F. Bidleman, *Environ. Sci. Technol.*, 1988, **22**, 361.
[72] S. J. Eisenreich, B. B. Looney and J. D. Thornton, *Environ. Sci. Technol.*, 1981, **15**, 30.

nature of the particle.[62,70,71] To a first approximation, chemical compounds with liquid-phase vapour pressures of $P_L < 10^{-6}$ Pa ($< 10^{-8}$ Torr) at the ambient atmospheric temperature are present in the particle phase, and those with values of $P_L > 10^{-2}$ Pa ($> 10^{-4}$ Torr) at the ambient atmospheric temperature are essentially totally in the gas phase.[71,72] Chemicals with intermediate values of P_L are present in both the gas and particle phases and are often termed semi-volatile organic compounds (SOCs).[71]

The liquid-phase vapor pressures, P_L, at 298 K of the PCBs, PCDDs and PCDFs decrease from 7 Pa (5×10^{-2} Torr) for biphenyl[65,73] to $\sim(1–3) \times 10^{-5}$ Pa [$(0.7–2.3) \times 10^{-7}$ Torr] for decachlorobiphenyl,[65,73] from 0.5 Pa (4×10^{-3} Torr) for dibenzo-*p*-dioxin[74] to $(3–5) \times 10^{-7}$ Pa [$(2–4) \times 10^{-9}$ Torr] for octachloro-dibenzo-*p*-dioxin,[74,75] and from 1.4 Pa (1×10^{-2} Torr) for dibenzofuran[74] to 1×10^{-7} Pa (8×10^{-10} Torr) for octachlorodibenzofuran.[74].

Based on theoretical considerations of gas–particle partitioning[70,71] and the results of several ambient air studies, the PCBs, PCDDs and PCDFs containing five or less chlorine atoms are present in the atmosphere at least partially in the gas phase at around room temperature.[14,22,30,38,39,47,51,62] The major PCB congeners observed in ambient air are those containing three and four chlorine atoms,[14–17,30,60–62] and several studies have reported that the gas-phase PCBs dominate over the particle-phase PCBs[14–17,30,41,45,60,61] (with, for example, the gas-phase PCBs being reported to comprise $>98\%$,[30] 83–97%[14] and 96%[41] of the total PCBs measured). Because of the variation of P_L with temperature, for a given particle surface area a decrease in ambient atmospheric temperature will increase the fraction of the SOC present in the particle phase (equation 1).

The phase (gas or particle) in which the PCBs, PCDDs and PCDFs occur in the atmosphere greatly affects their tropospheric removal processes and lifetimes. Analogous to other organic compounds, the PCDDs, PCDFs and PCBs in the atmosphere can be removed and/or transformed by a number of physical and chemical processes.[71,72,76,77] While the present chapter focuses on the chemical transformations of the PCBs, PCDDs and PCDFs, the physical removal processes are also discussed for completeness and to assess the relative importance of the various tropospheric removal processes.

3   Physical Removal Processes

Gas- and particle-phase PCBs, PCDDs and PCDFs can be removed from the troposphere by wet and dry deposition.[71,72,76,77] Wet deposition refers to the removal of the chemical (or particle-associated chemical) from the atmosphere by precipitation events (through the precipitation of rain, fog or snow to the Earth's surface), while dry deposition refers to the removal of the chemical or particle-

[73] W. T. Foreman and T. F. Bidleman, *J. Chromat.*, 1985, **330**, 203.

[74] B. F. Rordof, *Chemosphere*, 1989, **18**, 783.

[75] B. D. Eitzer and R. A. Hites, *Environ. Sci. Technol.*, 1988, **22**, 1362.

[76] R. Atkinson, in *Air Pollution, the Automobile, and Public Health*, ed. A. Y. Watson, R. R. Bates and D. Kennedy, National Academy Press, Washington, 1988, pp. 99–132.

[77] R. Atkinson, *Sci. Total Environ.*, 1991, **104**, 17.

associated chemical from the atmosphere to the Earth's surface by diffusion and/or sedimentation.[76]

Wet Deposition of PCBs, PCDDs and PCDFs

Wet deposition of a semi-volatile organic compound is characterized in terms of the overall washout ratio, W, which is given by

$$W = \text{(concentration of chemical in aqueous phase)/(concentration of chemical in air)} \tag{2}$$

$$= W_g(1 - \phi) + W_p\phi \tag{3}$$

where W_g is the washout ratio for the gas-phase chemical, W_p is the washout ratio for the particle-associated chemical, and ϕ is the fraction of the chemical which is particle associated (see equation 1). For a gas-phase chemical, wet deposition involves the incorporation of the chemical into rain-, fog or cloud-water, followed by precipitation to the Earth's surface, and is determined by the partitioning of the chemical between the air and aqueous phases. The gas-phase washout ratio, W_g is given by

$$W_g = C_w/C_a = RT/H \tag{4}$$

where C_w is the concentration of the chemical in the aqueous phase, C_a is the concentration of the chemical in air, R is the gas constant, T is the ambient atmospheric temperature (in K) and H is the Henry's law constant.[65,71,72] The flux of chemicals to the Earth's surface due to wet deposition, F_{wet}, is given by

$$F_{wet} = JWC_a \tag{5}$$

where J is the precipitation rate (for example, cm rain yr^{-1}).

Values of W_p for particle-associated chemicals depend on the type of precipitation event (warm or cold and in-cloud *versus* below-cloud precipitation) and the particle size,[71] and the magnitude of W_p appears to be in the range 10^2–10^6,[71,72,78] and generally in the range 10^4–10^6.[71,72] A value of $W_p = 10^5$ is calculated to result in a residence time for particles and particle-associated chemicals due to wet deposition in the well-mixed troposphere (with a scale height of 7 km) of ~20 days for a constant precipitation rate of 1 m yr^{-1}.

The Henry's law constants for the gas-phase PCBs are generally in the range 20–100 Pa m^3 mol^{-1} at 298 K and show little variation with the number of chlorine atoms, and the average value is $H = 48$ Pa m^3 mol^{-1} at 298 K.[65] This value of H corresponds to a washout ratio for PCBs of $W_g = 50$ at 298 K and an atmospheric residence time due to wet deposition of several tens of years (for removal from the well-mixed troposphere at an average temperature of 298 K

[78] M. P. Ligocki, C. Leuenberger and J. F. Pankow, *Atmos. Environ.*, 1985, **19**, 1619.

and a rainfall of 1 m yr^{-1}). Because $H = $ (vapor pressure/aqueous solubility), then H will decrease (and W_g will increase) with decreasing temperature, by a factor of ~ 10 for a temperature decrease of 25 K within the range 273–313 K.[79] Even though the majority of the PCBs are present in the lower troposphere in the gas phase (see above), the low washout ratios W_g for gas-phase PCBs results in precipitation scavenging of the particle-associated PCBs dominating over rain-out of gas-phase PCBs.[30,71,72]

For the PCDDs and PCDFs, Ligocki et al.[80] have reported a washout ratio for dibenzofuran of $W_g = 930 \pm 180$ at 281 K from ambient atmospheric measurements, while Eitzer and Hites[39] and Koester and Hites[81] derived gas-phase washout ratios for tetrachlorodibenzofuran, pentachlorodibenzo-p-dioxin and penta-chlorodibenzofuran of $(0.6–3) \times 10^4$. These washout ratios W_g derived from ambient measurements are reasonably consistent with the washout ratios calculated from the Henry's law constants estimated by Shui et al.[82] (which range between 0.1 and 15 Pa m^3 mol^{-1}, and decrease by a factor of ~ 1.6 per Cl atom added). The washout ratios W_g at 298 K estimated by Shui et al.[82] range from 200 for dibenzo-p-dioxin to 2×10^4 for heptachlorodibenzo-p-dioxin.

Ambient air studies[39,78,80,81] show that the particle-phase (and hence more chlorinated) PCDDs and PCDFs are more efficiently removed from the atmosphere by wet deposition than are the gas-phase PCDDs and PCDFs, analogous to the case for the PCBs.

Dry Deposition of PCBs, PCDDs and PCDFs

The flux of chemicals to the Earth's surface due to dry deposition, F_{dry}, is given by

$$F_{dry} = V_{dg}C_a + V_{dp}C_p \tag{6}$$

where V_{dg} and V_{dp} are the deposition velocities for the gas- and particle-phase chemicals, respectively (and hence V_{dp} is the deposition velocity for the particles with which the chemical is associated), and C_a and C_p are the atmospheric concentrations of the chemical in the gas and particle phases, respectively. The deposition velocities depend on the atmospheric stability, nature of the surface, nature of the chemical (for a gas-phase chemical) and (for a particle or particle-phase chemical) the size of the particle.[76] For particle deposition, the deposition velocity is a minimum for particles with mean diameter $\sim 0.3–0.5\ \mu m$, and increases with both increasing and decreasing particle size.[72,76] Ambient atmospheric deposition studies have concluded that the deposition of PCBs, PCDDs and PCDFs is largely due to deposition of the particle-phase compounds.[45,71,81] Because of the preferential deposition of larger particles, the average deposition velocity may well decrease with distance from the source region (for example, V_{dp} may be larger for particles in an urban area compared to

[79] L. P. Burkhard, D. E. Armstrong and A. W. Andren, Environ. Sci. Technol., 1985, 19, 590.
[80] M. P. Ligocki, C. Leuenberger and J. F. Pankow, Atmos. Environ., 1985, 19, 1609.
[81] C. J. Koester and R. A. Hites, Environ. Sci. Technol., 1992, 26, 1375.
[82] W. Y. Shui, W. Doucette, F. A. P. C. Gobas, A. Andren and D. Mackay, Environ. Sci. Technol., 1988, 22, 651.

those in more remote areas).[45] Values of V_{dp} of 0.1–1 cm s^{-1} for particle-phase organics (including PCBs),[71] 0.2 cm s^{-1} for particle-phase PCBs[18] and 0.06–0.6 (average 0.2) cm s^{-1} for particle-phase PCDDs and PCDFs[81] have been reported. With a value of $V_{dp} = 0.2$ cm s^{-1}, the calculated residence time of particles and particle-associated chemicals due to dry deposition in the well-mixed troposphere is ~ 30 days. It may be noted that modelling of the transport of particle-associated ^{210}Pb leads to an estimated residence time of particles in the atmosphere of ~ 5–15 days due to combined wet and dry deposition,[83] reasonably consistent with the above calculations.

The available data therefore indicate that deposition of PCBs, PCDDs and PCDFs from the atmosphere is dominated by wet and dry deposition of the particle-phase congeners.

4 Chemical Transformations

Gas- and particle-phase organic compounds can undergo chemical change *via* a number of routes.[76,77] For gas-phase chemicals, these involve photolysis and reaction with the hydroxyl (OH) radical, reaction with the nitrate (NO$_3$) radical and reaction with ozone (O$_3$).[76,77,84] The formation of OH radicals, NO$_3$ radicals and O$_3$ in the troposphere, and the requirements for photolysis to occur in the troposphere, are briefly discussed below, prior to presenting the experimental data for the PCBs, PCDDs and PCDFs for these processes.

Photolysis in the Troposphere

Absorption of solar radiation by molecular oxygen and ozone in the stratosphere limits the transmission of ultraviolet radiation into the troposphere to wavelengths ≥ 290 nm.[85] Any depletion of stratospheric ozone will lead to an increase in the intensity of UV radiation, and to shorter wavelength radiation, entering the troposphere.[86] For photolysis to occur, the chemical must absorb radiation in the 'actinic' region [between ~ 290 nm (the short-wavelength cut-off imposed by absorption of shorter wavelengths in the stratosphere) and ~ 800 nm (the longest wavelength that can lead to breakage of chemical bonds)]. Furthermore, having absorbed radiation, the chemical must undergo change. The photolysis rate, k_{phot}, is give by

$$k_{phot} = \int_{290\ nm}^{800\ nm} J_\lambda \sigma_\lambda \phi_\lambda \, d\lambda \tag{7}$$

where J_λ is the radiation flux at wavelength λ, σ_λ is the absorption cross-section at wavelength λ and ϕ_λ is the quantum yield for chemical change at wavelength λ.

[83] Y. J. Balkanski, D. J. Jacob, G. M. Gardner, W. C. Graustein and K. K. Turekian, *J. Geophys. Res.*, 1993, **98**, 20573.

[84] R. Atkinson, in *Volatile Organic Compounds in the Atmosphere* (*Issues in Environmental Science and Technology No. 4*), ed. R. E. Hester and R. M. Harrison, The Royal Society of Chemistry, Cambridge, 1995, pp. 65–89.

[85] G. Seckmeyer and R. L. McKenzie, *Nature*, 1992, **359**, 135.

[86] S. Madronich, R. L. McKenzie, M. M. Caldwell and L. O. Björn, *Ambio*, 1995, **24**, 143.

Ozone in the Troposphere

The presence of O_3 in the troposphere is due to downward transport from the stratosphere with dry deposition at the Earth's surface[84,87] and *in situ* chemical formation and destruction.[87,88] Mixing ratios of O_3 in the 'ccean' remote lower troposphere are in the range $(10–40) \times 10^{-9}$,[87,89] and increase with increasing altitude.[87] A 24 h average troposphere O_3 concentration of 7×10^{11} molecule cm^{-3} appears reasonable for lifetime calculations.

OH Radicals in the Troposphere

The presence of O_3 in the troposphere leads to the formation of OH radicals, by the photolysis of O_3 at wavelengths 290–320 nm to form the electronically excited oxygen atom, $O(^1D)$, which either reacts with water vapor or is deactivated by reaction with O_2 and N_2 to the ground state oxygen atom, (O^3P), which then rapidly recombines with O_2 to reform O_3.[84,90]

$$O_3 + h\nu \rightarrow O_2 + O(^1D)\ (\lambda < 320\,nm) \tag{8}$$

$$O(^1D) + H_2O \rightarrow 2\,OH \tag{9}$$

$$O(^1D) + M \rightarrow O(^3P) + M\ (M = N_2, O_2) \tag{10}$$

A diurnally, seasonally and annually averaged global tropospheric concentration of the OH radical has been derived from the emissions, atmosphere concentrations and OH radical reaction rate constant for methyl chloroform (CH_3CCl_3), resulting in a 24 h average OH radical concentration of 9.7×10^5 molecule cm^{-3}.[91] Because of the fairly long lifetimes of the PCBs, PCDDs and PCDFs (see below), for lifetime calculations this globally averaged tropospheric OH radical concentration is more relevant than ground-level direct measurements of OH radical concentrations.

NO_3 Radicals in the Troposphere

Emissions of NO from combustion processes and soils and *in situ* formation of NO from lightning are followed by reactions leading to the formation of the NO_3 radical:[84,90]

$$NO + O_3 \rightarrow NO_2 + O_2 \tag{11}$$

$$NO_2 + O_3 \rightarrow NO_3 + O_2 \tag{12}$$

[87] J. A. Logan, *J. Geophys. Res.*, 1985, **90**, 10463.
[88] G. P. Ayers, S. A. Penkett, R. W. Gillett, B. Bandy, I. E. Galbally, C. P. Meyer, C. M. Elsworth, S. T. Bentley and B. W. Forgan, *Nature*, 1992, **360**, 446.
[89] S. J. Oltmans and H. Levy II, *Atmos. Environ.*, 1994, **28**, 9.
[90] R. Atkinson, D. L. Baulch, R. A. Cox, R. F. Hampson, Jr., J. A. Kerr and J. Troe, *J. Phys. Chem. Ref. Data*, 1992, **21**, 1125.
[91] R. G. Prinn, R. F. Weiss, B. R. Miller, J. Huang, F. N. Alyea, D. M. Cunnold, P. J. Fraser, D. E. Hartley and P. G. Simmonds, *Science*, 1995, **269**, 187.

Because the NO_3 radical photolyses rapidly,[90] NO_3 radical concentrations are low during daylight hours, but can become elevated during nighttime. Measured ground-level NO_3 radical concentrations range up to 1×10^{10} molecule cm^{-3}, and a 12 h nighttime average concentration of $\sim 5 \times 10^8$ molecule cm^{-3}, uncertain by a factor of ~ 10, has been proposed.[92]

5   Transformations of Gas-phase PCBs, PCDDs and PCDFs

Reactions with O_3

To date, the kinetics of the gas-phase reactions of O_3 with biphenyl, 2,2'-dichlorobiphenyl, dibenzo-*p*-dioxin, 1-chlorodibenzo-*p*-dioxin and dibenzofuran have been studied at room temperature.[93–95] In all cases, no reactions were observed and the measured upper limits to the rate constants are given in Table 1. The observed lack of reaction of O_3 with gas-phase PCBs, PCDDs and PCDFs is expected,[77] based on literature data for other aromatic compounds,[96,97] and it is anticipated that the room temperature rate constants for the gas-phase reactions of O_3 with PCBs, PCDDs and PCDFs are all $<2 \times 10^{-20}$ cm^3 molecule^{-1} s^{-1}.

Reactions with the NO_3 Radical

The gas-phase reactions of the NO_3 radical with biphenyl, 2,2'-dichlorobiphenyl, dibenzo-*p*-dioxin, 1-chlorodibenzo-*p*-dioxin and dibenzofuran have been studied at room temperature.[92,93,95,98,99] While no reactions of the NO_3 radical with biphenyl, 2,2'-dichlorobiphenyl or dibenzofuran were observed,[92,93,95,98,99] dibenzo-*p*-dioxin and 1-chlorodibenzo-*p*-dioxin were observed to react, with the measured reaction rate constants increasing with increasing NO_2 concentration.[95,99] As discussed in detail by Atkinson[92] and Kwok *et al.*,[99] the reactions of the NO_3 radical with the PCBs, PCDDs and PCDFs appear to proceed by initial addition of the NO_3 radical to form an NO_3 adduct, which either thermally decomposes back to reactants or reacts with O_2 and/or NO_2 (Scheme 1). The measured rate constant, k_{obs}, is then given by

$$k_{obs} = k_a(k_c[NO_2] + k_d[O_2])/(k_b + k_c[NO_2] + k_d[O_2]) \qquad (13)$$

The experimental data for dibenzo-*p*-dioxin, 1-chlorodibenzo-*p*-dioxin[95,99] and polycyclic aromatic hydrocarbons[92] show that under the experimental conditions employed, $k_c[NO_2] > k_d[O_2]$ and that equation (13) simplifies to

$$k_{obs} = k_a k_c[NO_2]/(k_b + k_c[NO_2]) \qquad (14)$$

[92] R. Atkinson, *J. Phys. Chem. Ref. Data*, 1991, **20**, 459.
[93] R. Atkinson, S. M. Aschmann and J. N. Pitts, Jr., *Environ. Sci. Technol.*, 1984, **18**, 110.
[94] E. S. C. Kwok, J. Arey and R. Atkinson, *Environ. Sci. Technol.*, 1994, **28**, 528.
[95] E. S. C. Kwok, R. Atkinson and J. Arey, *Environ. Sci. Technol.*, 1995, **29**, 1591.
[96] R. Atkinson and W. P. L. Carter, *Chem. Rev.*, 1984, **84**, 437.
[97] R. Atkinson, *J. Phys. Chem. Ref. Data*, 1994, **Monograph 2**, 1.
[98] R. Atkinson, J. Arey, B. Zielinska and S. M. Aschmann, *Environ. Sci. Technol.*, 1987, **21**, 1014.
[99] E. S. C. Kwok, R. Atkinson and J. Arey, *Int. J. Chem. Kinet.*, 1994, **26**, 511.

Table 1 Rate constants k for the gas-phase reactions of O_3 with PCBs, PCDDs and PCDFs at room temperature

Aromatic	$10^{20} \times k$ (cm^3 molecule^{-1} s^{-1})	T(K)	Reference
PCBs			
Biphenyl	<20	294 ± 1	Atkinson et al.[93]
2,2'-Dichlorobiphenyl	<2	297 ± 2	Kwok et al.[95]
PCDDs			
Dibenzo-*p*-dioxin	<5	297 ± 2	Kwok et al.[94]
1-Chlorodibenzo-*p*-dioxin	<7	297 ± 2	Kwok et al.[95]
PCDFs			
Dibenzofuran	<8	297 ± 2	Kwok et al.[94]

Scheme 1

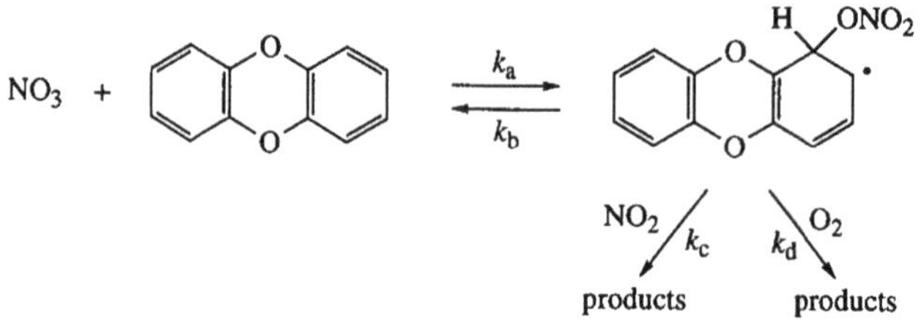

In addition, a direct reaction pathway (for example, involving H-atom abstraction) should be considered (although the experimental data do not show any evidence for this pathway, nor is it expected[92]):

$$NO_3 + \text{aromatics} \rightarrow \text{products} \tag{15}$$

where the rate constant for reaction (15) is referred to as k_{abs}.

For biphenyl, 2,2'-dichlorobiphenyl and dibenzofuran, no reactions were observed[93,95,98,99] and only upper limits to the rate constants k_{abs} and rate constant ratios $k_a k_c / k_b$ are available, and these are given in Table 2. In contrast, reactions of dibenzo-*p*-dioxin and 1-chlorodibenzo-*p*-dioxin with the NO_3 radical were observed,[95,99] with the measured reaction rate constants k_{obs} increasing with increasing NO_2 concentration in accord with equations (13) and (14). The rate constant ratios $k_a k_c / k_b$ obtained from the experimental data are given in Table 2. The measured rate constants in the absence of NO_2 allow upper limits to the rate constants k_{abs} to be obtained,[95,99] and these are also given in Table 2. For dibenzo-*p*-dioxin, an upper limit to the rate constant ratio k_d / k_c has also been derived (Table 2),[99] showing that the reaction of the NO_3-dibenzo-*p*-dioxin adduct with NO_2 dominates over reaction with O_2 for NO_2 mixing ratios $\geq 8 \times 10^{-8}$ (and possibly much lower since only an upper limit to the rate constant ratio k_d / k_c was obtained). Rate constants k_a for the initial addition of the NO_3 radical to dibenzo-*p*-dioxin[95] and 1-chlorodibenzo-*p*-dioxin can also be

Table 2 Rate constants k_{abs} and rate constant ratios $k_a k_c/k_b$ and k_d/k_c for the gas-phase reactions of the NO_3 radical with PCBs, PCDDs and PCDFs at room temperature

Aromatic	k_{abs} (cm^3 molecule^{-1} s^{-1})	$k_a k_c/k_b$ (cm^6 molecule^{-2} s^{-1})	k_d/k_c	T(K)	Reference
PCBs					
Biphenyl	$<6 \times 10^{-16}$	$<5 \times 10^{-30}$		298 ± 2	Atkinson et al.[92,98]
2,2'-Dichlorobiphenyl	$<8 \times 10^{-16}$	$<3 \times 10^{-30}$ [a]		297 ± 2	Kwok et al.[95]
PCDDs					
Dibenzo-p-dioxin	$<8 \times 10^{-15}$	3.9×10^{-27} [b]	$<4 \times 10^{-7}$	297 ± 2	Kwok et al.[99]
1-Chlorodibenzo-p-dioxin	$<1.6 \times 10^{-14}$	5.7×10^{-28} [b,c]		297 ± 2	Kwok et al.[95]
PCDFs					
Dibenzofuran	$<1.6 \times 10^{-15}$	$<7 \times 10^{-30}$		297 ± 2	Kwok et al.[99]

[a]Rate constant ratio cited in Kwok et al.[95] was erroneously a factor of 10 low.
[b]Uncertain to a factor of ~2.
[c]Reanalysis of the experimental data of Kwok et al.[95]

derived, with values at $297 \pm 2 K$ of 6.8×10^{-14} cm^3 molecule^{-1} s^{-1} and 2.7×10^{-13} cm^3 molecule^{-1} s^{-1}, respectively, both with overall uncertainties of a factor of ~ 2.

The rate constants presented in Table 2 suggest that the NO$_3$ radical reactions with the gas-phase PCBs and PCDFs have upper limits to the rate constants for any reaction of $k_{abs} < 10^{-15}$ cm^3 molecule^{-1} s^{-1} (NO$_2$ concentrations in the troposphere are sufficiently low that the contribution of any NO$_2$-dependent reaction is encompassed within the upper limit to the bimolecular reaction rate constant k_{abs}). While dibenzo-p-dioxin and 1-chlorodibenzo-p-dioxin react with the NO$_3$ radical by the NO$_2$-dependent mechanism shown in Scheme 1 and equation (13), the NO$_2$ concentrations in the troposphere are sufficiently low that the effective bimolecular rate constants, $k_a(k_c[NO_2] + k_d[O_2])/k_b$, are below the upper limits to the rate constants k_{abs}.

Reactions with the OH Radical

The experimentally measured rate constants for the gas-phase reactions of the OH radical with PCBs, PCDDs and PCDFs are given in Table 3. The majority of the rate constants measured have been for the PCBs, with the measurements of Anderson and Hites[104,105] providing rate constants for a number of trichloro-, tetrachloro- and pentachlorobiphenyls. It should be noted that Anderson and Hites[104,105] measured rate constants for the PCBs at elevated temperatures in the range 322–366 K, with the minimum temperature at which measurements were made being 322 or 323 K apart from 4,4'-dichlorobiphenyl (329 K) and the three pentachlorobiphenyls (343 K). Arrhenius expressions were obtained from these data and used to calculate the rate constants at 298 K by extrapolation.[104,105] The rate constants measured by Anderson and Hites[104,105] in the temperature range 322–366 K all show a decrease in rate constant with decreasing temperature, corresponding to values of B in the Arrhenius expression $k = A \exp(-B/T)$ ranging from 20 ± 150 K for biphenyl to 1950 ± 470 K for 2,2',3,5',6-pentachlorobiphenyl (where the indicated errors are one standard deviation).

Table 3 shows that the room temperature rate constants for biphenyl reported by Zetzsch,[100] Atkinson et al.,[93,101] Atkinson and Aschmann,[101,102] Klöpffer et al.[101,103] and Anderson and Hites[104] are in good agreement. Furthermore, the room temperature rate constants reported by Atkinson and Aschmann[101,102] and Anderson and Hites[104] for the monochlorobiphenyls are in excellent agreement, as are those of Kwok et al.[95] and Anderson and Hites[104] for 2,2'-dichlorobiphenyl. Table 3 shows that in general the rate constants for the biphenyls decrease with increasing chlorination, and this also appears to be the case for the PCDDs based on the rate constants for dibenzo-p-dioxin and 1-chlorodibenzo-p-dioxin. The data for the PCBs also show that for a given degree of chlorination, the rate constant depends on the substitution pattern around the aromatic rings (Table 3).

As also concluded previously,[77] reaction with the OH radical may be the only important tropospheric loss process for the gas-phase PCBs, PCDDs and PCDFs. It is therefore necessary to know the rate constants for the gas-phase

Table 3 Rate constants for the gas-phase reactions of the OH radical with PCBs, PCDDs and PCDFs at room temperature

Aromatic	$(10^{12} \times k\,cm^3\ molecule^{-1}\ s^{-1})$	$T(K)$	Reference
PCBs			
Biphenyl	5.8 ± 0.8	296	Zetzsch[100]
	7.61 ± 0.67	294 ± 1	Atkinson et al.[93,101]
	8.32 ± 0.75	295 ± 1	Atkinson and Aschmann[101,102]
	8.0	300	Klöpffer et al.[101,103]
	$6.7^{+1.0}_{-0.8}$	298	Anderson and Hites[104]
2-Chlorobiphenyl	2.82 ± 0.38	295 ± 1	Atkinson and Aschmann[101,102]
	$2.7^{+0.7}_{-0.6}$	298	Anderson and Hites[104]
3-Chlorobiphenyl	5.28 ± 0.82	295 ± 1	Atkinson and Aschmann[101,102]
	$5.0^{+0.7}_{-0.6}$	298	Anderson and Hites[104]
4-Chlorobiphenyl	3.86 ± 0.67	295 ± 1	Atkinson and Aschmann[101,102]
	$3.4^{+0.6}_{-0.5}$	298	Anderson and Hites[104]
2,2'-Dichlorobiphenyl	2.0 ± 0.5	297 ± 2	Kwok et al.[95]
	2.2 ± 0.5^a	298	Anderson and Hites[104]
2,4-Dichlorobiphenyl	$2.6^{+0.7}_{-0.5}\,^a$	298	Anderson and Hites[104]
3,3'-Dichlorobiphenyl	4.1 ± 1.2	297 ± 2	Kwok et al.[95]
3,5-Dichlorobiphenyl	4.2 ± 1.1	297 ± 2	Kwok et al.[95]
4,4'-Dichlorobiphenyl	2.0 ± 0.5^a	298	Anderson and Hites[105]
2',3,4-Trichlorobiphenyl	$1.0^{+0.4}_{-0.2}\,^a$	298	Anderson and Hites[104]
2,4,4'-Trichlorobiphenyl	1.1 ± 0.3^a	298	Anderson and Hites[104]
2,4,5-Trichlorobiphenyl	1.3 ± 0.2^a	298	Anderson and Hites[104]
2,4',5-Trichlorobiphenyl	$1.2^{+0.3}_{-0.2}\,^a$	298	Anderson and Hites[104]
2,2',3,5'-Tetrachlorobiphenyl	$0.8^{+0.4}_{-0.2}\,^a$	298	Anderson and Hites[104]
2,2',4,4'-Tetrachlorobiphenyl	$1.0^{+0.5}_{-0.3}\,^a$	298	Anderson and Hites[104]
2,2',3,5',6-Pentachlorobiphenyl	$0.4^{+0.3}_{-0.2}\,^a$	298	Anderson and Hites[104]
2,3,3',4',6-Pentachlorobiphenyl	$0.6^{+0.5}_{-0.3}\,^a$	298	Anderson and Hites[104]
2,3,4,5,6-Pentachlorobiphenyl	$0.9^{+0.9}_{-0.5}\,^a$	298	Anderson and Hites[104]
PCDDs			
Dibenzo-*p*-dioxin	14.8 ± 3.5	297 ± 2	Kwok et al.[94]
1-Chlorodibenzo-*p*-dioxin	4.7 ± 1.6	297 ± 2	Kwok et al.[95]
PCDFs			
Dibenzofuran	3.9 ± 0.9	297 ± 2	Kwok et al.[94]

^{a}Rate constants calculated at 298 K from measurements carried out at elevated temperatures in the range 322–366 K.

65

reactions of the OH radical with the various PCB, PCDD and PCDF congeners in order to estimate the lifetimes of these compounds. However, because of experimental limitations, rate constants have been measured for only a few of the large number of PCB, PCDD and PCDF congeners, and then only for the more volatile and less chlorinated congeners (Table 3).

Atkinson[77] and Kwok et al.[95] have proposed and discussed methods for the estimation of OH radical reaction rate constants for the PCBs, PCDDs and PCDFs, based on the correlation of the rate constants for OH radical addition to aromatic rings, k^{OH}, with the sum of the electrophilic substituent constants, $\Sigma\sigma^+$.[77,100] Based on a review of the literature rate constants for OH radical addition to a wide range of aromatic compounds, Atkinson[77] derived the correlation,

$$\log k^{OH} \ (cm^3 \ molecule^{-1} \ s^{-1}) = -11.71 - 1.34\Sigma\sigma^+ \tag{16}$$

and no new data have been reported which change this correlation. $\Sigma\sigma^+$ is calculated[100] by assuming that (a) steric hindrance can be neglected, (b) $\Sigma\sigma^+$ is the sum of the electrophilic substituent constants of all of the substituent groups attached to the aromatic rings, (c) the OH radical adds to the position yielding the most negative value of $\Sigma\sigma^+$ and (d) if all positions are occupied, the *ipso* position is treated as a *meta* position. Atkinson[77] used the rate constants for the reactions of the OH radical with biphenyl and the monochlorobiphenyls and the literature values of the electrophilic substituent constants for the Cl and phenyl groups[106] to derive electrophilic substituent constants for the $C_6H_{5-x}Cl_x(x = 1–4)$ groups, and the relevant electrophilic substituent constants are given in Table 4. A comparison of estimated (using equation (16) and the electrophilic substituent constants given in Table 4) and measured (Table 3) room temperature rate constants k^{OH} for the PCB congeners for which rate constants have been measured is shown in Figure 2, where the solid line denotes perfect agreement between the estimated and experimentally measured rate constants and the dashed lines denote disagreement by a factor of 2. Figure 2 shows that the estimated rate constants are in good agreement with the experimental data, with the estimated values possibly being low by $\sim 25\%$. This generally good agreement for the PCBs containing up to five chlorine atoms indicates that this estimation method can be used to reliably calculate (to within a factor of ~ 2) the room temperature rate constants k^{OH} for the PCB congeners for which experimental measurements are not yet available. The estimated room temperature rate constants for PCBs containing up to six chlorine atoms are given in Table 5; as noted above, for a given degree of chlorination the rate constant depends on the Cl atom substitution pattern around the aromatic rings.

[100] C. Zetzsch, presented at the 15th Informal Conference on Photochemistry, Stanford, CA, June 27–July 1, 1982.
[101] R. Atkinson, J. Phys. Chem. Ref. Data, 1989, **Monograph 1**, 1.
[102] R. Atkinson and S. M. Aschmann, Environ. Sci. Technol., 1985, **19**, 462.
[103] W. Klöpffer, R. Frank, E.-G. Kohl and F. Haag, Chem.-Ztg., 1986, **110**, 57.
[104] P. N. Anderson and R. A. Hites, Environ. Sci. Technol., 1996, **30**, 1756.
[105] P. N. Anderson and R. A. Hites, Environ. Sci. Technol., 1996, **30**, 301.
[106] H. C. Brown and Y. Okamoto, J. Am. Chem. Soc., 1958, **80**, 4979.

Table 4 Electrophilic substituent constants $\sigma_{o,p}^+$ and σ_m^+ used in the estimation of OH radical reaction rate constants for PCBs, PCDDs and PCDFs

Substituent	$\sigma_{o,p}^+$	σ_m^+	Reference
Cl	0.114	0.399	Brown and Okamoto[106]
C_6H_5	−0.179	0.109	Brown and Okamoto[106]
C_6H_4Cl	−0.039	0.207	Atkinson[77]
$C_6H_3Cl_2$	0.101	0.305	Atkinson[77]
$C_6H_2Cl_3$	0.241	0.403	Atkinson[77]
C_6HCl_4	0.381	0.501	Atkinson[77]
C_6Cl_5	0.521	0.599	Atkinson[77]
C_6H_5O	−0.30	0.00	Kwok et al.[95]
C_6H_4ClO	−0.20[a]	0.07[a]	
$C_6H_3Cl_2O$	−0.10[a]	0.14[a]	
$C_6H_2Cl_3O$	0.00[a]	0.21[a]	
C_6HCl_4O	0.10[a]	0.28[a]	

[a] Derived from σ^+ (C_6H_5O) assuming that the effect of increasing Cl atom substitution on the C_6H_5O group is 70% of the effect of increasing Cl atom substitution on the C_6H_5 group.

Figure 2 Comparison of the experimentally measured and estimated room temperature rate constants k_{exp}^{OH} and k_{calc}^{OH}, respectively, for the gas-phase reactions of the OH radical with PCBs. Solid line denotes perfect agreement of k_{exp}^{OH} and k_{calc}^{OH}, while the dashed lines denote disagreement by a factor of 2.0. The rate constants are tabulated in Table 3: ○, Anderson and Hites;[104,105] ●, Atkinson et al.,[93,101] Atkinson and Aschmann[101,102] and Kwok et al.;[95] △, Zetzsch;[100] ▽, Klöpffer et al.[101,103]

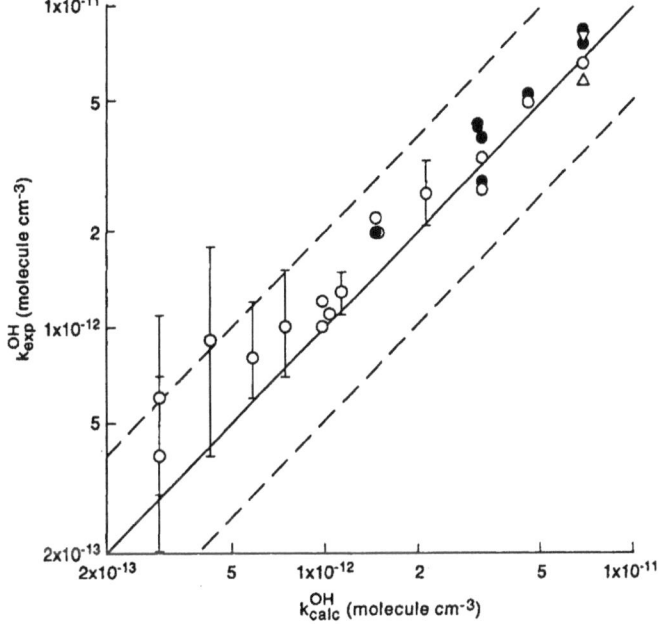

Analogous estimation methods have been proposed for the PCDDs and PCDFs,[77,95] with the most recent study of Kwok et al.[95] proposing that the PCDDs be viewed as two benzene rings with Cl and phenoxy ($C_6H_{5-x}Cl_xO$, with $x = 0$–4) substituent groups, and that the PCDFs be viewed as two benzene rings with Cl, phenoxy ($C_6H_{5-x}Cl_xO$, with $x = 0$–4) and phenyl ($C_6H_{5-x}Cl_x$, with $x = 0$–4) substituent groups. Kwok et al.[95] used the measured room temperature rate constants for dibenzo-*p*-dioxin, dibenzofuran and diphenyl ether ($C_6H_5OC_6H_5$) to derive electrophilic substituent constants for the C_6H_5O group and then

Table 5 Estimated room temperature rate constants k^{OH} for the gas-phase reactions of the OH radical with gas-phase PCBs, PCDDs and PCDFs

No. Cl atoms	$10^{12} \times k^{OH}$ (cm^3 molecule^{-1} s^{-1})		
	PCBs	PCDDs[a]	PCDFs[a]
0	6.8	9.8 (14.8 ± 3.5)[b]	7.0 (3.9 ± 0.9)[b]
1	3.2–4.6	6.4 (4.7 ± 1.6)[b]	4.3–4.4
2	1.4–3.1	2.7–4.2	1.8–2.8
3	1.0–2.1	1.7–2.7	1.0–1.7
4	0.35–1.7	0.7–1.7	0.4–1.0
5	0.3–0.9	0.45–0.75	0.25–0.45
6	0.16–0.5	0.2–0.4	0.1–0.2

[a]Using the σ^+ (C$_6$H$_{5-x}$Cl$_x$) values given in Table 4.
[b]Experimental values (see Figure 2 for a comparison between the measured and calculated rate constants for the PCBs).

calculated room temperature rate constants for PCDDs and PCDFs containing up to five chlorine atoms, assuming that the electrophilic substituent constants of C$_6$H$_{5-x}$Cl$_x$O groups ($x = 1$–4) were identical to those for the C$_6$H$_5$O group. However, it is expected that the electrophilic substituent constants for the C$_6$H$_{5-x}$Cl$_x$O group will depend on the value of x, and the rate constants calculated by Kwok et al.[95] for the PCDDs and PCDFs were probably too high, being increasingly overestimated as the degree of chlorination increases. The phenoxy group may exhibit a somewhat lesser effect of Cl substitution on the electrophilic substituent constant than is the case for the phenyl group (because the Cl atoms are further from the attachment point to the aromatic ring). Electrophilic substituent constants for the C$_6$H$_{5-x}$Cl$_x$O group have been derived assuming that the decrease in σ^+ ($\Delta\sigma^+$) per additional Cl atom on the phenoxy group is ~ 0.7 of that for the phenyl group, and these values of σ^+ are given in Table 4. It should, of course, be noted that this is all very empirical and that the rate constants estimated for the PCDDs and PCDFs (Table 5) are quite uncertain, the more so the higher the degree of chlorination. For example, if the decrease in σ^+ ($\Delta\sigma^+$) per additional Cl atom on the phenoxy group is the same as that for the phenyl group, then the calculated rate constants for the PCDDs and PCDFs are lower than those given in Table 5 by a factor which increases with the number of chlorine atoms (with this factor being 1.1–1.3 for the PCDDs and PCDFs containing two chlorine atoms and 1.4–1.7 for those containing five chlorine atoms).

Photolysis

While there have been a number of studies concerning the absorption cross-sections, photolysis rates and photolysis quantum yields of PCBs,[107–109] PCDDs[107,109–117]

[107] D. Dulin, H. Drossman and T. Mill, Environ. Sci. Technol., 1986, **20**, 72.
[108] M. Koshioka, J. Kanazawa, H. Iizuka and T. Murai, Bull. Environ. Contam. Toxicol., 1987, **38**, 409.
[109] M. Koshioka, M. Ishizaka, T. Yamada, J. Kanazawa and T. Murai, J. Pestic. Sci., 1990, **15**, 439.
[110] G. G. Choudhry and G. R. B. Webster, Chemosphere, 1985, **14**, 9.
[111] R. T. Podoll, H. M. Jaber and T. Mill, Environ. Sci. Technol., 1986, **20**, 490.
[112] G. G. Choudhry and G. R. B. Webster, Chemosphere, 1986, **15**, 1935.

and PCDFs[109] in the solution phase, there have been few studies dealing with the absorption cross-sections or photolysis rates of chlorinated biphenyls, dibenzo-*p*-dioxins or dibenzofurans in the gas phase.[118-120] Berman *et al.*[119] measured the absorption spectra and cross-sections of the first absorption band of 3,3',4,4'-tetrachlorobiphenyl at 250, 300 and 350°C, and observed that the long-wavelength threshold of this band was red-shifted with increasing temperature, from 302 nm at 250°C to 306 nm at 350°C.[119] Extrapolation of these long-wavelength thresholds[119] to lower temperatures results in a long-wavelength threshold for 3,3',4,4'-tetrachlorobiphenyl of 293 nm at 298 K, with marginal overlap of the absorption spectrum with the tropospheric solar spectrum. In the study of Orth *et al.*,[118] 2,3,7,8-tetrachlorodibenzo-*p*-dioxin was photolysed at 150°C over the wavelength range ~250–340 nm. While photolysis was observed,[118] this could have been due to photolysis at wavelengths ≤ 290 nm (shorter than those in the troposphere) and the effect of elevated temperature may also have been significant. Funk *et al.*[120] measured the absorption spectra and cross-sections of three PCDDs and two PCDFs at elevated temperatures (120–300°C), and observed the peak wavelengths and approximate long-wavelength thresholds of the first absorption band to be at 303 nm and ~340 nm for both 2,3,7,8-tetrachlorodibenzo-*p*-dioxin and 1,2,3,7,8-pentachlorodibenzo-*p*-dioxin, at 305 nm and ~360 nm for octachlorodibenzo-*p*-dioxin, at 300 nm and ~340 nm for 2,3,7,8-tetrachlorodibenzofuran and at 333 nm and ~365 nm for octachlorodibenzofuran, with no evidence for shifts of the peak wavelength with temperature over the limited ranges employed.[120] It should be noted that the long-wavelength threshold for absorption of dibenzofuran in the gas phase at room temperature is at ~299–300 nm,[121] indicating that the chlorinated dibenzofurans will absorb radiation within the actinic region in the troposphere.

6   Transformations of Particle-phase PCBs, PCDDs and PCDFs

Few direct data are available concerning the chemical and photochemical degradations of PCBs, PCDDs and PCDFs present in the particle phase. Tysklind and Rappe[122] irradiated fly-ashes containing PCDDs and PCDFs from incineration plants with artificial UV light (at wavelengths ≥ 300 nm) for up to 288 h. After 288 h the fly-ash samples showed ~10–15% losses of the tetrachloro- to octachlorodibenzo-*p*-dioxins and ~0–20% losses of the tetrachloro-to octachlorodibenzofurans.[122] Similar results were obtained for sunlight

[113] M. Koshioka, T. Yamada, J. Kanazawa and T. Murai, *J. Pestic. Sci.*, 1989, **14**, 181.

[114] M. Koshioka, M. Ishizaka, T. Yamada, J. Kanazawa and T. Murai, *J. Pestic. Sci.*, 1990, **15**, 39.

[115] L.-S. Hung and L. L. Ingram, Jr., *Bull Environ. Contam. Toxicol.*, 1990, **44**, 380.

[116] B. Guan and P. Wan, *J. Chem. Soc., Chem. Commun.*, 1993, 409.

[117] B. Guan and P. Wan, *J. Photochem. Photobiol., A*, 1994, **80**, 199.

[118] R. G. Orth, C. Ritchie and F. Hileman, *Chemosphere*, 1989, **18**, 1275.

[119] J. M. Berman, J. L. Graham and B. Dellinger, *J. Photochem. Photobiol., A*, 1992, **68**, 353.

[120] D. J. Funk, R. C. Oldenborg, D.-P. Dayton, J. P. Lacosse, J. A. Draves and T. J. Logan, *Appl. Spectrosc.*, 1995, **49**, 105.

[121] A. R. Auty, A. C. Jones and D. Phillips, *Chem. Phys. Lett.*, 1984, **112**, 529.

[122] M. Tysklind and C. Rappe, *Chemosphere*, 1991, **23**, 1365.

irradiation for 28 h.[122] Koester and Hites[123] used a rotary photoreactor to irradiate fly-ashes containing PCDDs and PCDFs at wavelengths >295 nm (with the light intensities at 295–305 nm and 310–315 nm being higher than natural sunlight at 50° zenith angle). Within the experimental uncertainties of ±35%, Koester and Hites[123] observed no photodegradation of the tetrachloro- to octachlorodibenzo-p-dioxins or dibenzofurans upon irradiating four fly-ashes (ranging in colour from black to grey to yellow) for up to 200 h.[123] Furthermore, irradiation of pre-cleaned black fly-ash spiked with selected tetrachloro- to hexachlorodibenzo-p-dioxins and dibenzofurans for 100 h led to no photodegradation of the spiked PCDDs or PCDFs within the experimental uncertainties.[123] These data[122,123] indicate that photolysis of fly-ashes containing PCDDs and PCDFs is of negligible or minor importance in the atmosphere, and this conclusion may also hold for particle-phase PCDDs, PCDFs and PCBs in general.

7 Tropospheric Lifetimes of PCBs, PCDDs and PCDFs

Gas-phase PCBs, PCDDs and PCDFs

The little information available (see above) suggests that photolysis of gas-phase PCBs in the troposphere will be negligible for those PCBs with ≤ 4 chlorine atoms, and this may also be the case for the more chlorinated PCBs. However, it appears that the PCDDs and PCDFs will absorb radiation in the actinic region in the troposphere, and hence the importance of photolysis will depend on the photodissociation quantum yields in the gas phase, which are, however, not presently known.

The measured rate constants, or upper limits thereof, for the gas-phase reactions of OH radicals, NO_3 radicals and O_3 can be combined with the estimated ambient tropospheric concentrations of OH radicals, NO_3 radicals and O_3 to calculate the tropospheric lifetimes of the gas-phase PCBs, PCDDs and PCDFs due to each of these reactions. The lifetime, τ_x, for reaction with species X is given by $\tau_x = (k_x[X])^{-1}$. Ambient concentrations (molecule cm^{-3}) of: OH radicals, a 24 h, seasonal, annual and global tropospheric average of 9.7×10^5;[91] NO_3 radicals, a 12 h nighttime average of 5×10^8;[92] and O_3, a 24 h average of 7×10^{11} (30×10^{-9} mixing ratio)[87] are used to calculate lifetimes due to these gas-phase reactions.

The upper limits to the O_3 reaction rate constants given in Table 1, and the expected upper limit to the rate constants for the gas-phase reactions of O_3 with PCB, PCDD and PCDF congeners of $<2 \times 10^{-20}$ cm^3 molecule^{-1} s^{-1} (see above), lead to a calculated lower limit to the tropospheric lifetimes of PCB, PCDD and PCDF congeners due to gas-phase reaction with O_3 of >2 years. Similarly, the upper limit to the measured bimolecular rate constants k_{abs} for reactions of the NO_3 radical with PCBs, PCDDs and PCDFs under tropospheric conditions, of $k_{abs} < 1 \times 10^{-15}$ cm^3 molecule^{-1} s^{-1}, leads to a lower limit to the tropospheric lifetimes of PCB, PCDD and PCDF congeners due to gas-phase reaction with the NO_3 radical of >45 days.

In contrast, reactions of the PCBs, PCDDs and PCDFs with the OH radical

---

[123] C. J. Koester and R. A. Hites, *Environ. Sci. Technol.*, 1992, **26**, 502.

Table 6 Calculated
tropospheric lifetimes of
gas-phase PCBs, PCDDs
and PCDFs due to
reaction with the OH
radical

No. Cl atoms	Lifetime $(d)^a$		
	PCBs	PCDDs	PCDFs
0	1.8	1.2 $(0.8)^b$	1.7 $(3.1)^b$
1	2.6–3.7	1.9 $(2.5)^b$	2.7
2	3.8–9	2.8–4.4	4.3–6.6
3	5.7–12	4.4–7	7–12
4	7–34	7–17	12–30
5	13–40	16–27	27–48
6	24–75	30–60	60–120

^{a}Calculated using the estimated OH radical reaction rate constants given in Table 5 and
assuming a 24 h average OH radical concentration of 9.7×10^5 molecule cm^{-3}.[91]
^{b}Using the experimentally measured reaction rate constants.

lead to calculated lifetimes, using the measured and estimated OH radical
reaction rate constants given in Tables 3 and 5, ranging from 1.7 days for
biphenyl, 0.8 days for dibenzo-*p*-dioxin and 3 days for dibenzofuran to ∼13–40
days for the pentachlorobiphenyls, ∼16–27 days for the pentachlorodibenzo-*p*-
dioxins and ∼27–48 days for the pentachlorodibenzofurans. The calculated
lifetimes for the OH radical reactions are given in Table 6. Within the PCBs,
PCDDs and PCDFs the calculated lifetimes increase with increasing chlorination,
by a factor of ∼1.7 per extra chlorine atom for each class of organochlorine
compound. The lifetimes for the PCB congeners are expected to be reasonably
reliable, because of the existence of measured rate constants for a fairly large
number of PCBs, including pentachlorobiphenyls (Table 3). In contrast, apart
from dibenzo-*p*-dioxin, 1-chlorodibenzo-*p*-dioxin and dibenzofuran, the estimated
lifetimes for the OH radical reactions with the PCDD and PCDF congeners are
quite uncertain, and as discussed above the uncertainties increase with increasing
chlorination. It should be recognized that in all cases the calculated lifetimes are
inversely proportional to the tropospheric OH radical concentration assumed.

Particle-phase PCBs, PCDDs and PCDFs

As discussed above, the PCBs, PCDDs and PCDFs present in the atmosphere in
the particle phase will be removed from the troposphere by wet and dry
deposition of the particles, and potentially also by photolysis and/or reaction of
the particle-phase PCBs, PCDDs and PCDFs. At the present time, insufficient
data are available to assess the importance of photolysis and/or chemical
reactions of the particle-phase congeners, although the studies of Tysklind and
Rappe[122] and Koester and Hites[123] suggest that photolysis is not important.
Upper limits to the lifetimes of the particle-phase PCBs, PCDDs and PCDFs are
given by those for wet and dry deposition of the particles, with these lifetimes
depending on a number of factors, as discussed above. Based on the approximate
calculations given above for the lifetimes of particles due to wet and dry
deposition, and the data of Balkanski *et al.*[83] for aerosol-phase ^{210}Pb, the
lifetimes of particle-associated PCBs, PCDDs and PCDFs due solely to wet and
dry deposition are expected to be in the range ∼5–30 days, depending on particle

size, rainfall and how vertically well-mixed are the PCBs, PCDDs and PCDFs in the troposphere.

8 Conclusions

In the atmosphere, the PCBs, PCDDs and PCDFs are distributed between the gas and particle phases, with the more chlorinated congeners partitioning more into the particle phase. In general, the PCBs, PCDDs and PCDFs containing ≤ 5 chlorine atoms are present largely in the gas phase, with the congeners containing ≥ 6 chlorine atoms being present mainly in the particle phase (although this gas/particle partitioning is a function of the atmospheric temperature, with lower temperatures favouring partitioning into the particle phase, and on the particle loading in the atmosphere). The available laboratory and ambient air data indicate that the dominant tropospheric loss processes for the PCBs PCDDs and PCDFs are wet and dry deposition for those PCBs, PCDDs and PCDFs present in the atmosphere in the particle phase, and reaction with the OH radical for those PCBs, PCDDs and PCDFs present in the atmosphere in the gas phase. However, photolysis of gas-phase PCDDs and PCDFs and reaction of gas-phase PCDDs with the NO$_3$ radical cannot presently be ruled out as being of importance as PCDD and/or PCDF loss processes in the troposphere. The estimated lifetimes of the gas-phase PCBs, PCDDs and PCDFs increase with increasing chlorination, from ~ 0.8–3 days for biphenyl, dibenzo-*p*-dioxin and dibenzofuran to ~ 15–50 days for the pentachloro-biphenyls, dibenzo-*p*-dioxins and -dibenzofurans (with the calculated lifetimes being inversely dependent on the OH radical concentration assumed). The lifetimes for the particle-phase and more highly chlorinated PCBs, PCDDs and PCDFs appear to be determined by wet and dry deposition of the 'host' particles, with estimated lifetimes of ~ 5–30 days, depending on a number of factors including the frequency of precipitation events. The lifetimes of the chlorinated biphenyls, dibenzo-*p*-dioxins and dibenzofurans are > 1 day and therefore allow (especially for the congeners containing ≥ 3 chlorine atoms) long-range transport of these organochlorine compounds into remote areas of the world.

There are clearly several areas of significant uncertainty, including the role of photolysis in both the gas and particle phases and the identities and formation yields of the products of the gas-phase OH radical reactions.

9 Acknowledgements

Drs. R. A. Hites and P. N. Anderson are gratefully thanked for communicating their rate constant data for the gas-phase reactions of the OH radical with PCBs prior to publication, and Dr. J. Arey is thanked for helpful discussions.

Human Toxicology of Chlorinated Organic Micropollutants

STEPHEN SAFE

1 Introduction

Chlorine has been extensively used as an industrial compound for the synthesis of innumerable commercial products, including bleaches, solvents, disinfectants, monomers for plastics, pharmaceuticals and pesticides. Moreover, chlorine gas is widely used as a disinfectant for water purification systems. Despite the acknowledged societal contributions of chlorinated chemicals, there recently has been considerable media attention on proposals by some environmental groups and scientists to ban the industrial use of chlorine.[1-3] The idea of banning a natural elemental chemical which plays an integral role in the function of natural ecosystems and biosystems has also been challenged by scientists and scientific organizations.[4-6] The origins of chlorine chemophobia are derived primarily from the environmental impact and hypothesized adverse human impacts of one major subclass of chlorinated compounds, namely the persistent chlorinated hydrocarbons which have been identified as widespread environmental contaminants (Figure 1).[7-8] Included in this group of compounds are chemicals which have been produced for specific applications, namely polychlorinated biphenyls (PCBs), naphthalenes (PCNs) and terphenyls (PCTs); organochlorine pesticides, such as DDT and related metabolites; toxaphene; and chlorinated cyclodiene-derived compounds (*e.g.* dieldrin, endrin and endosulfan). In addition, combustion of organic material which contains a source of chlorine also results in the formation of polychlorinated dibenzo-*p*-dioxins (PCDDs) and dibenzofurans (PCDFs). PCDDs and PCDFs as well as PCBs have been detected as

[1] I. Amato, *Science*, 1993, **261**, 152.
[2] J. Thornton, *The Product is the Poison: The Case for A Chlorine Phase-Out. A Greenpeace Report*, Greenpeace, USA, 1991.
[3] P. L. deFur, *Health Environ. Digest.*, 1995, **8**, 11.
[4] I. Amato, *Garbage*, 1994, **6**, 30.
[5] M. H. Karol, *Fundam. Appl. Toxicol.*, 1995, **24**, 1.
[6] G. W. Gribble, *Environ. Sci. Technol.*, 1994, **28**, 310A.
[7] D. W. Kuehl and B. Butterworth, *Chemosphere*, 1994, **29**, 523.
[8] S. Tanabe, N. Kannan, A. A. Subramanian, S. Watanabe and R. Tatsukawa, *Environ. Pollut.*, 1987, **47**, 147.

by-products in the combustion of municipal, hospital and various organic wastes, in car exhausts and cigarette smoke.[9-11] PCDDs are also by-products formed during the production of chlorinated phenols and their derived products, including the herbicides 2,4-dichlorophenoxyacetic acid (2,4-D) and 2,4,5-trichlorophenoxyacetic acid (2,4,5-T).[12-16] PCDDs and PCDFs have also been detected as by-products in the combustion of wood and coal, and the relative global mass contributions of various industrial-related *versus* 'natural' (*e.g.* forest fires, volcanic eruptions) sources of PCDDs and PCDFs have not been unequivocally determined.[10,16] A major problem associated with human exposures to PCDDs and PCDFs is the high toxicity of 2,3,7,8-tetrachlorodibenzo-*p*-dioxin (TCDD) and related 2,3,7,8-substituted PCDDs and PCDFs.[17,18]

The environmental problems associated with organohalogen compounds are primarily due to their physicochemical properties, namely chemical stability and lipophilicity. After their release into the environment, organochlorine pollutants are stable and resistant to chemical and biodegradation and preferentially bioaccumulate in the food chain.[16] Thus residues of PCBs, PCDDs, PCDFs, DDT and other organochlorine pesticides are routinely detected in fish, wildlife and human adipose tissue, serum and milk.[7,8,17,18] The relative levels of these compounds are variable; however, in most locations, *p, p'*-DDE (a DDT metabolite) and PCBs are the dominant contaminants and are present at high parts per billion to low parts per million levels, whereas PCDDs and PCDFs are usually detected at the parts per trillion to parts per billion levels.[19,20] Concentrations of other halogenated hydrocarbon pollutants tend to be between those observed for PCBs/DDE and PCDD/Fs. The discovery of DDT and related compounds and PCBs as global environmental contaminants resulted in the restricted use or banning of these chemicals in most developed nations; however, DDT is still used as an insecticide in several less developed or underdeveloped countries. The application and use of most other organochlorine compounds has also been restricted or banned. Not surprisingly, levels of these contaminants have decreased in most locations,[21-24] and, in the Great Lakes

9   S. Marklund, E. Wilkström, G. Lofvenius, I. Fangmark and C. Rappe, *Chemosphere*, 1994, **28**, 1895.
10 H. Fiedler, *Chemosphere*, 1996, **32**, 55.
11 K. Olie, P. L. Vermeulen and O. Hutzinger, *Chemosphere*, 1977, **6**, 455.
12 O. Hutzinger and H. Fiedler, *Chemosphere*, 1993, **27**, 121.
13 H. Hagenmaier and H. Brunner, *Chemosphere*, 1987, **20**, 2425.
14 W. Christmann, K. D. Kloppel, H. Partscht and W. Rotard, *Chemosphere*, 1989, **18**, 861.
15 A. Norstrom, C. Rappe, R. Lindahl and H. R. Buser, *Scand. J. Work. Environ. Health*, 1979, **5**, 375.
16 S. Safe, *Environ. Carcin. Ecotox. Rev.*, 1991, **9**, 261.
17 S. Safe, *Annu. Rev. Pharmacol. Toxicol.*, 1986, **26**, 371.
18 J. A. Goldstein and S. Safe, in *Halogenated Biphenyls, Naphthalenes, Dibenzodioxins and Related Compounds*, ed. R. D. Kimbrough and A. A. Jensen, Elsevier, Amsterdam, 1989, p. 239.
19 K. Ballschmiter, H. Buchert and S. Bihler, *Fresenius' Z. Anal. Chem.*, 1981, **306**, 323.
20 S. Tanabe, H. Iwata and R. Tatsukawa, *Sci. Total Environ.*, 1994, **154**, 163.
21 R. Turle, R. J. Norstrom and B. Collins, *Chemosphere*, 1991, **22**, 201.
22 C. J. Schmitt, J. L. Zajicek and P. H. Peterman, *Arch. Environ. Contam. Toxicol.*, 1990, **19**, 748.
23 C. A. Lake, J. L. Lake, R. Haebler, R. McKinney, W. S. Boothman and S. S. Sadove, *Arch. Environ. Contam. Toxicol.*, 1995, **29**, 128.
24 L. L. Williams, J. P. Giesy, D. A. Verbrugge, S. Jurzysta, G. Heinz and K. L. Stromborg, *Arch. Environ. Contam. Toxicol.*, 1995, **29**, 52.

Figure 1

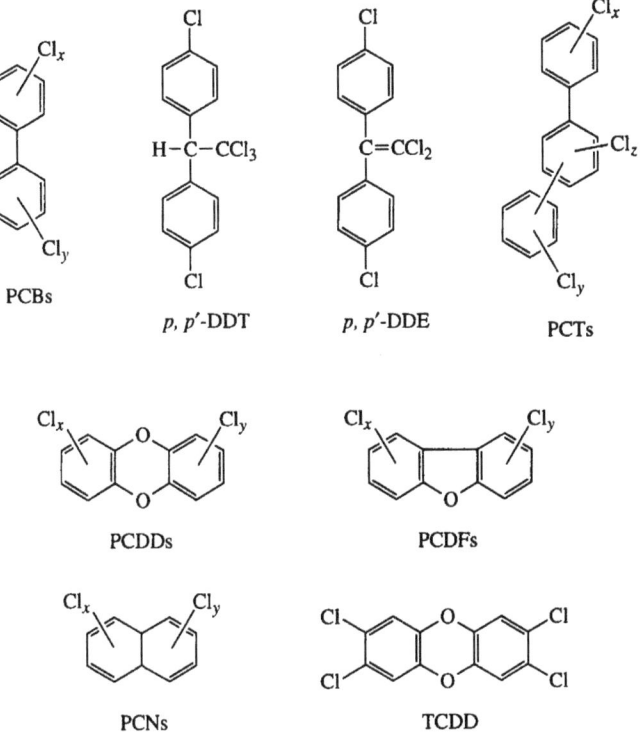

region, lower levels of some organochlorine pollutants are paralleled by increased wildlife (birds) reproduction.[25]

2 Human Toxicology of Organochlorine Pollutants

Background

Human exposure to halogenated hydrocarbons has primarily occurred through three different exposure scenarios, namely (1) background exposure in which the major route is *via* consumption of contaminated food; (2) accidental exposures; and (3) occupational exposure during production or use of organochlorine compounds. The highest exposure group is usually industrial workers, and their levels of exposure are dependent on their length of employment in contaminated areas. Most accidental exposures occur over a short time period; however, depending on exposure levels, these individuals may represent high, medium or low exposure groups. For example, after an explosion in a chemical factory in Seveso, Italy, released a 'toxic cloud' containing high levels of TCDD, individuals residing in the most contaminated areas (zone A) had tissue or serum levels of TCDD which were higher than most occupationally exposed workers.[26,27] Most

[25] J. P. Giesy, J. P. Ludwig and D. E. Tillitt, *Environ. Sci. Technol.*, 1994, **28**, 128A.

[26] G. Reggiani, *J. Toxicol. Environ. Health*, 1980, **6**, 27.

[27] P. Mocarelli, L. L. Needham, A. Marocchi, D. G. Patterson, Jr., P. Brambilla, P. M. Gerthoux, L. Meazza and V. Carreri, *J. Toxicol. Environ. Health*, 1991, **32**, 357.

of the adverse human health effects resulting from exposures to halogenated aromatic pollutants are derived from the high-dose occupational/accidental exposures. The potential adverse effects of lower level environmental exposures are inferred from the higher exposures and results from laboratory animal studies. This chapter will focus on the human toxicology of PCDDs, PCDFs and PCBs.

3 PCDDs/PCDFs

Introduction

The biochemical and toxic effects of PCDDs and PCDFs in laboratory animals and humans have been extensively investigated.[17,18,28-30] TCDD is the most toxic member of this class of compounds and has been used as a prototype for investigating responses elicited by PCDDs and PCDFs and for determining their mechanism of action. TCDD induces a diverse spectrum of toxic and biochemical responses in laboratory animals and cells in culture, and these include a wasting syndrome, immune suppression, hepatotoxicity and porphyria, reproductive and developmental toxicity, carcinogenicity, tumour promotion activity, disruption of endocrine response pathways, and induction of both phase 1 (CYP1A genes and related enzyme activities) and phase 2 drug-metabolizing enzymes. The effects of 2,3,7,8-substituted PCDD/Fs in laboratory animals and humans are age-, sex-, strain- and species-dependent, and in both *in vivo* and *in vitro* studies the response to PCDD/Fs are highly tissue- and cell-type-specific. The mechanism of action of TCDD and related compounds has been extensively investigated using induction of CYP1A1 gene expression as a model.[29,31,32] The uptake of TCDD into target cells is followed by initial binding to the cytosolic aryl hydrocarbon receptor (AhR) protein, which rapidly accumulates in the nucleus as a heterodimer. The nuclear AhR complex contains the AhR and AhR nuclear translocator (Arnt) proteins which subsequently interact with *cis*-acting dioxin responsive elements (DREs) to effect transactivation. It is generally accepted that most of the biochemical and toxic responses caused by TCDD and related compounds are AhR mediated. The AhR has been identified in most target tissues in laboratory animals, mammalian cells in culture and human tissues/organs.[33-35]

Carcinogenesis

The carcinogenic effects of PCDDs and PCDFs have been extensively investigated in several highly exposed groups, including industrial workers, herbicide applicators, individuals poisoned in the Yusho and Yu-Cheng incidents in Japan

[28] A. Poland and J. C. Knutson, *Annu. Rev. Pharmacol. Toxicol.*, 1982, **22**, 517.

[29] J. P. Whitlock, Jr., *Chem. Res. Toxicol.*, 1993, **6**, 754.

[30] G. W. Lucier, C. J. Portier and M. A. Gallo, *Environ. Health Perspect.*, 1993, **101**, 36.

[31] S. Safe, *Pharmacol. Therap.*, 1995, **67**, 247.

[32] H. I. Swanson and C. A. Bradfield, *Pharmacogenetics*, 1993, **3**, 213.

[33] J. P. Landers and N. J. Bunce, *Biochem. J.*, 1991, **276**, 273.

[34] A. B. Okey, D. S. Riddick and P. A. Harper, *Toxicol. Lett.*, 1994, **70**, 1.

[35] S. Safe, in *Hazards, Decontamination and Replacement of PCBs*, ed. J. P. Crin, Plenum Press, New York, 1988, p. 51.

	Response	Comments
Table 1 Carcinogenic and non-carcinogenic effects associated with human exposures to PCDDs/PCDFs	Soft tissue sarcoma	Observed in some but not all studies
	Non-Hodgkins lymphoma	Observed in some but not all studies
	Increased cancer mortality	Increased in some studies, particularly in individuals with high exposure and/or increased latency times
	Increased respiratory cancers, liver cancer, stomach cancer	Observed only in a few studies
	Decreased breast and endometrial cancer	Observed in one study
	Chloracne and related dermal lesions	Consistently observed in some individuals with high exposure
	Induction of liver or serum enzyme levels	One or more indicators observed in several studies
	Effects on thyroid function, gastro-intestinal disorders, neurotoxicity, pulmonary and cardiovascular function deficits, diabetes and immune suppresion	Observed in one or more studies

and Taiwan, respectively, and individuals exposed in the Seveso accident in Italy.[36-45] The utility and significance of several epidemiology studies are variable due to a number of factors, which include limited numbers of exposed individuals, small numbers of tumours, exposure to chemicals other than PCDDs and PCDFs (*e.g.* phenoxy herbicides, PCBs and other industrial chemicals), problems in tumour detection and classification and minimal data on actual levels of exposure to PCDD/Fs. Table 1 summarizes results of studies reported by several groups on the association between exposure to PCDD/Fs and development of cancer.

A number of cohort studies on workers exposed to PCDD/Fs during the manufacture of chlorinated herbicides or phenols have been reported. Analysis of the cancer mortality data from 5712 workers in herbicide manufacturing plants in

[36] M. A. Fingerhut, W. E. Halperin, D. A. Marlow, L. A. Piacitelli, P. A. Honchar, M. H. Sweeney, A. L. Greife, P. A. Dill, K. Steenland and A. J. Surunda, *New Engl. J. Med.*, 1991, **324**, 212.

[37] A. Manz, J. Berger, J. H. Dwyer, D. Fleschjanys, S. Nagel and H. Waltsgott, *Lancet*, 1991, **338**, 959.

[38] A. Zober, P. Messerer and P. Huber, *Int. Arch. Occup. Environ. Health*, 1990, **62**, 139.

[39] R. Sarcci, M. Kogevinas, P. A. Bertazzi, B. H. B. Demesquita, D. Coggon, L. M. Green, T. Kauppinen, K. A. Labbe, M. Littorin, E. Lynge, J. D. Mathews, J. Neuberger, J. Osman, N. Pearce and R. Winkelmann, *Lancet*, 1991, **338**, 1027.

[40] H. B. Bueno de Mesquita, G. Doornbos, D. A. Van der Kuip, M. Kogevinas and R. Winkelmann, *Am. J. Ind. Med.*, 1993, **23**, 289.

[41] E. Lynge, *Br. J. Cancer*, 1985, **52**, 259.

[42] E. Lynge, *Cancer Causes Control*, 1993, **4**, 261.

[43] D. Coggon, B. Pannett and P. Winter, *Br. J. Ind. Med.*, 1991, **48**, 173.

[44] P. A. Bertazzi, A. C. Pesatori, D. Consonni, A. Tironi, M. T. Landi and C. Zocchetti, *Epidemiology*, 1993, **4**, 398.

[45] K. Wiklund and L. E. Holm, *J. Natl. Cancer Inst.*, 1986, **76**, 229.

the United States showed that death from all cancers was slightly elevated (SMR = 115).[36] Mortality from several cancers which were associated with TCDD exposure in some other studies (*e.g.* stomach and nasal cancer, non-Hodgkin's lymphoma and soft tissue sarcoma) were not increased; however, in workers with ≥ 1 year of exposure and ≥ 20 years of latency, the SMR values were significantly increased for all cancers (SMR = 146), soft tissue sarcoma (SMR = 929) and for cancers of the respiratory system (SMR = 142). The authors concluded that 'This study of mortality among workers with occupational exposure to TCDD does not confirm the high relative risks reported for many cancers in previous studies. ... we cannot exclude the possible contributions of factors such as smoking and occupational exposure to other chemicals'. In a study of 18 910 production workers and herbicide sprayers from 10 countries, there was no significant increase in cancer mortality.[39] There was an increased risk for soft tissue sarcoma and several other cancers occurring in the group with a high latency (10–19 years from first exposure). However, the excess cancer 'does not seem to be specifically associated with those herbicides probably contaminated by TCDD'.[39]

Cancer mortality in 1583 workers employed in herbicide manufacturing in Germany was increased in the whole group (SMR = 124–139) and the SMR value increased to 182–187 for men with 20 or more years employment in the plant.[37] Interestingly, breast cancer mortality was increased in exposed women (SMR = 215) whereas breast cancer incidence was decreased in women exposed to TCDD in the Seveso incident.[44] This latter observation is consistent with laboratory studies which demonstrate that TCDD is a potent antiestrogen in human breast cancer cells and inhibits mammary tumour formation and growth in rodent models.[31] No deaths due to soft tissue sarcoma were found in the German workers; however, there was increased mortality from non-Hodgkin's lymphoma. The third major industrial study examined BASF employees who were exposed to TCDD during an accident in 1953.[38] Many of these individuals developed chloracne, which is a hallmark of TCDD exposure. The SMR for all cancers was 117 and, for workers with chloracne, the SMR was higher (139) but not significantly elevated. The authors concluded that there was not a strong association between TCDD exposure and increased incidence of cancer; however, the size of the exposed cohort may be too small to detect an increase.

An industrial accident in a chemical plant resulted in the release of TCDD and exposure of the population in Seveso, Italy.[26,44] The cancer incidence of individuals living in zones A, B and R (in order of decreasing levels of contamination) have been examined. The SMRs for all malignancies were 100, 80 and 90 for females and 70, 100 and 90 for males in zones A, B and R, respectively. Interpretation of the results of this study is difficult because the number of exposed individuals in zone A < zone B < zone R, and there was not a consistent exposure-dependent increase in any tumour. In zone R, the incidence of non-Hodgkin's lymphoma and soft tissue sarcoma was increased; in zone B, hepatobiliary cancer was elevated. Breast and endometrial cancer were decreased in most areas, and this is consistent with the antiestrogenic activity of TCDD and related compounds.[31]

Several case-control studies examined the association between phenoxy herbicide exposure and non-Hodgkin's lymphoma and soft tissue sarcoma.[46-60]

Studies in Sweden[46-51] correlated increased incidence of non-Hodgkin's lymphoma and soft tissue sarcoma with exposure to phenoxy herbicides and chlorinated phenols, and this was also reported in other studies. In contrast, other case control studies[52-57] did not observe these correlations, and this issue has been the subject of controversy in the scientific literature. Two recent nested case-control studies reported that excess risk of soft tissue sarcoma was associated with exposure to any phenoxy herbicide or to 2,4-dichloro- and 2,4,5-trichloro-phenoxyacetic acid and 4-chloro-2-methylphenoxyacetic acid; risks for non-Hodgkin's lymphoma were 'generally weaker than those found in the study on sarcoma'.[61] These data do not necessarily implicate TCDD as an etiologic agent since exposure to herbicides which do not contain TCDD as a contaminant were also associated with increased risk from soft tissue sarcoma. Thus, because of the variations between studies, both the cohort and case-control studies do not unequivocally show that TCDD in association with herbicides induces specific cancers in humans. This is an area of research which has been extensively reviewed[62-64] and there is considerable controversy regarding the interpretation of results and study designs.[65,66]

Non-cancer Endpoints

One of the major hallmarks of acute exposure to TCDD and related compounds is the development of chloracne and other dermal problems which have been identified in highly exposed workers.[67,68] In animal studies, TCDD can cause

[46] L. Hardell and A. Sandström, *Br. J. Cancer*, 1979, **39**, 711.

[47] L. Hardell and M. Eriksson, *Cancer*, 1988, **62**, 652.

[48] L. Hardell, M. Eriksson, P. Lenner and E. Lundgren, *Br. J. Cancer*, 1981, **43**, 169.

[49] M. Eriksson, L. Hardell and H. O. Adami, *J. Natl. Cancer Inst.*, 1990, **82**, 486.

[50] M. Eriksson, L. Hardell, N. O. Berg, T. Moller and O. Axelson, *Br. J. Ind. Med.*, 1981, **38**, 27.

[51] L. Hardell, M. Eriksson and A. Degerman, *Cancer Res.*, 1994, **54**, 2386.

[52] A. H. Smith, D. O. Fisher, N. Pearce and C. J. Chapman, *Arch. Environ. Health*, 1982, **37**, 197.

[53] A. H. Smith, D. O. Fisher, H. J. Giles, *et al.*, *Chemosphere*, 1983, **12**, 565.

[54] A. H. Smith, N. E. Pearce, D. O. Fisher, H. J. Giles, C. A. Teague and J. K. Howard, *J. Natl. Cancer Inst.*, 1984, **73**, 1111.

[55] A. H. Smith and N. E. Pearce, *Chemosphere*, 1986, **15**, 1795.

[56] N. E. Pearce, *Br. J. Ind. Med.*, 1989, **46**, 143.

[57] N. E. Pearce, A. H. Smith and D. O. Fisher, *Am. J. Epidemiol.*, 1985, **121**, 225.

[58] J. S. Woods, L. Polissar, R. K. Severson, L. S. Heuser and B. G. Kulander, *J. Natl. Cancer Inst.*, 1987, **78**, 899.

[59] S. H. Zahm, D. D. Weisenburger, P. A. Babbitt, R. C. Saal, J. B. Vaught, K. P. Cantor and A. Blair, *Epidemiology*, 1990, **1**, 349.

[60] K. P. Cantor, A. Blair, G. Everett, *et al.*, *Cancer Res.*, 1992, **52**, 2447.

[61] M. Kogevinas, T. Kauppinen, R. Winkelmann, H. Becher, P. A. Bertazzi, H. B. Bueno-de-Mesquita, D. Coggon, L. Green, E. Johnson and M. Littorin, *Epidemiology*, 1995, **6**, 396.

[62] United States Environmental Protection Agency, *Health Assessment Document for 2,3,7,8-Tetrachlorodibenzo-p-dioxin (TCDD) and Related Compounds*, 111, 7, USEPA, Washington, 1994.

[63] E. S. Johnson, *CRC. Crit. Rev. Toxicol.*, 1992, **21**, 451.

[64] D. E. Lilienfeld and M. A. Gallo, *Epidemiol. Rev.*, 1989, **11**, 28.

[65] L. Hardell, *Br. J. Cancer*, 1993, **67**, 1154.

[66] J. G. Smith and A. J. Christophers, *Br. J. Cancer*, 1993, **67**, 1156.

[67] K. D. Crow, *New Sci.*, 1978, **78**, 78.

[68] M. Moses and P. G. Prioleau, *J. Am. Acad. Dermatol.*, 1985, **12**, 497.

liver damage and both induction and repression of diverse enzyme activities. A number of studies have investigated the effects of exposure to PCDD/Fs on various indicators of hepatic effects, and these include elevation of serum enzymes such as γ-glutamyl transferase (GGT), aspartate amino transferase (AST), alanine aminotransferase, D-glucaric acid secretion, increased serum triglycerides, CYP1A2 induction and porphyria.[69-75] Unfortunately, most studies provide only limited data on one or more of these parameters; however, the results suggest that there may be some reversible effects in liver. A recent study[76] investigated the induction of caffeine metabolism, a CYP1A2-dependent response, in workers with highly elevated serum TCDD levels (Table 2). The results showed that induction of CYP1A2 was not associated with elevated levels of TCDD; however, if the exposed and control groups were sorted into smoking and non-smoking groups, the results showed that significantly elevated caffeine metabolite ratios were observed in the former group. Thus, for induction of CYP1A2, the compounds in cigarette smoke were more effective as inducers of CYP1A2 than the high levels of TCDD in the worker cohort. This suggests that cigarette smoking and the relatively high intake of other AhR agonists in various foods may be confounding factors in assessing adverse impacts of TCDD related compounds.[77] It has been suggested that risk assessment of current dietary levels of TCDD and related compounds (exodioxins) should also take into account the contributions of 'natural dioxins'.[77]

Several studies have examined other possible adverse effects associated with exposure to PCDD/Fs, and these included alteration of thyroid function, gastrointestinal disorders, reproductive problems, neurotoxicity, immunotoxicity, pulmonary and cardiovascular effects.[71,78-88] Many of these effects have been

[69] P. Mocarelli, A. Marocchi, P. Brambilla, P. Gerthoux, D. S. Young and N. Mantel, *J. Am. Med. Assoc.*, 1986, **256**, 2687.

[70] F. Caramaschi, G. del Corno, C. Favaretti, S. E. Giambelluca, E. Montesarchio and G. M. Fara, *Int. J. Epidemiol.*, 1981, **10**, 135.

[71] M. Moses, R. Lilis, K. D. Crow, J. Thorenton, A. Fischbein, H. A. Anderson and I. J. Selikoff, *Am. J. Ind. Med.*, 1984, **5**, 161.

[72] G. Ideo, G. Bellati, A. Bellobuono and L. Bissanti, *Environ. Health Perspect.*, 1985, **60**, 151.

[73] L. Jirasek, J. Kalensky and K. Kubec, *Cesk. Dermatol.*, 1973, **48**, 306.

[74] L. Jirasek, J. Kalensky, K. Kubec, J. Pazderova and E. Lukas, *Cesk. Dermatol.*, 1974, **49**, 145.

[75] G. M. Calvert, M. H. Sweeney, M. A. Fingerhut, R. W. Hornung and W. E. Halperin, *Am. J. Ind. Med.*, 1994, **25**, 559.

[76] W. E. Halperin, W. Kalow, M. H. Sweeney, B. K. Tang, M. Fingerhut, B. Timpkins and K. Wille, *Occup. Environ. Med.*, 1995, **52**, 86.

[77] S. Safe, *J. Anim. Sci.*, 1996, in press.

[78] R. R. Suskind and V. S. Hertzberg, *J. Am. Med. Assoc.*, 1984, **251**, 2372.

[79] W. Wolf, J. Michalek, J. Miner, L. Needham and D. Patterson, Jr., *Organohalogen Compd.*, 1992, **10**, 279.

[80] W. Wolfe, J. Michalek, J. Miner, A. Rahe, L. Needham and D. Patterson, Jr., *Organohalogen Compd.*, 1992, **10**, 283.

[81] R. E. Hoffman, P. A. Stehr-Green, K. B. Webb, R. G. Evans, A. P. Knutsen, W. F. Schramm, J. L. Staake, B. B. Gibson and K. K. Steinberg, *J. Am. Med. Assoc.*, 1986, **255**, 2031.

[82] K. B. Webb, R. G. Evans, A. P. Knutsen, S. T. Roodman, D. W. Roberts, W. F. Schramm, B. B. Gibson, J. S. Andrews, Jr., L. L. Needham and D. G. Patterson, *J. Toxicol. Environ. Health*, 1989, **28**, 183.

[83] R. Alderfer, M. Sweeney, M. Fingerhut, R. Hornung, K. Wille and A. Fidler, *Chemosphere*, 1992, **25**, 247.

Table 2 Induction of CYP1A2-dependent caffeine metabolism in herbicide workers with elevated levels of TCDD[76]

Effects of TCDD on CMR	Controls (125)	Workers (58)
TCDD levels (pg g^{-1})	6.7 ± 4.3	15.7 ± 322
Caffeine metabolite ratio (CMR)	7.3 ± 4.4	7.7 ± 5.0

Effects of Smoking on CMR		TCDD (pg g^{-1})	CMR
Smokers	workers (13)	217.2 ± 471.5	10.9 ± 3.4[a]
	control (31)	5.3 ± 2.1	11.6 ± 5.5[a]
Non-smokers	workers (45)	139.6 ± 269.5	6.8 ± 5.0
	control (94)	7.2 ± 4.7	5.9 ± 2.8

[a]Significantly higher than non-smokers ($p < 0.05$).

observed in laboratory animals exposed to TCDD; however, these responses were not consistently observed (or measured) in exposed human populations. The association between increased risk for diabetes and elevated fasting serum glucose levels and TCDD exposure was reported in the Ranch Hand studies which examined highly exposed Vietnam veterans,[81] and this is an area which should also be further investigated. In summary, high levels of adult exposure to PCDD/Fs result in some adverse effects which appear to be reversible; therefore, it is unlikely that low level dietary exposure to these compounds represents a risk to human health.

4 PCBs

Introduction

PCBs are industrial compounds which were widely used as organic diluents, plasticizers, pesticide extenders, adhesives, dust-reducing agents, cutting oils, flame retardants, heat transfer fluids, dielectric fluids for transformers and capacitors, hydraulic lubricants, sealants and in carbonless copy paper. PCBs have entered the environment *via* multiple pathways and residues of these compounds have been identified in air, water, wildlife and human adipose tissue, serum and milk.[89-91] Like many other aromatic hydrocarbons, PCBs are highly lipophilic and chemically stable and this has contributed to their environmental persistence and bioconcentration in the food chain. PCBs were originally

[84] J. A. Zack and R. R. Suskind, *J. Occup. Med.*, 1980, **22**, 11.

[85] P. A. Mertazzi, C. Zocchetti, A. C. Pesatori, S. Guercilena, M. Sanarico and L. Radice, *Am. J. Epidemiol.*, 1989, **129**, 1187.

[86] M. Hatch, in *Public Health Risks of the Dioxins*, ed. W. W. Lowrance, Kaufmann, California, 1984, p. 255.

[87] G. M. Calvert, R. W. Hornung, M. H. Sweeney, M. A. Fingerhut and W. E. Halperin, *J. Am. Med. Assoc.*, 1992, **267**, 2209.

[88] G. M. Calvert, M. H. Sweeney, J. A. Morris, M. A. Fingerhut, R. W. Hornung and W. E. Halperin, *Am. Rev. Respir. Dis.*, 1991, **144**, 1302.

[89] S. Safe, *CRC. Crit. Rev. Toxicol.*, 1984, **12**, 319.

[90] S. Safe, *CRC. Crit. Rev. Toxicol.*, 1990, **21**, 51.

[91] S. Safe, *CRC. Crit. Rev. Toxicol.*, 1994, **24**,. 87.

manufactured and sold according to their degree of chlorination, and therefore PCB residues in the environment are derived from multiple PCB formulations. There are 209 possible individual PCB congeners and at least 132 of these compounds have been identified in the commercial mixtures.[92] The PCB composition of environmental samples is highly variable and dependent on the specific sample (*e.g.* air, water or biota) and the relative PCB inputs from local or regional sources of these contaminants.[93,94] PCB residues from wildlife and humans resemble, in part, the gas chromatographic pattern observed for highly chlorinated Aroclor 1260 (60% by weight) due to the preferential environmental degradation of lower chlorinated compounds.

The toxic and biochemical effects of PCB mixtures and individual compounds have been extensively investigated in laboratory animals and mammalian cells in culture.[89-91]. The pattern of PCB (mixture)-induced responses is highly complex since the mixtures contain structurally diverse compounds which elicit a wide spectrum of responses. For example, many of the effects of commercial PCB mixtures resemble those described for TCDD and related compounds. These activities are primarily due to the non-*ortho* or coplanar PCB congeners (3,3',4,4',5-pentaCB and 3,3',4,4',5,5'-hexaCB) or their mono-*ortho*-substituted analogues (2',3,4,4',5-pentaCB, 2,3,4,4',5-pentaCB, 2,3,3',4,4'-pentaCB, 2,3',4,4',5-pentaCB, 2,3,3',4,4'5-hexaCB, 2,3',4,4',5,5'-hexaCB, 2,3,3',4,4',5'-hexaCB and 2,3,3',4,4',5,5'-heptaCB). All of these compounds bind to the Ah receptor and elicit Ah receptor-mediated responses; however, the mono-*ortho*-substituted congeners also resemble phenobarbital (PB) as inducers of hepatic drug-metabolizing enzymes and are called 'mixed-type' inducers.[95,96] Diverse structural classes of PCBs, typified by 2,2',4,4',5,5'-hexaCB, resemble PB as an inducer of drug-metabolizing enzymes and as tumour promoters in rodent models.[96,97,98] More recent studies have identified '*ortho*-substituted' PCBs which are neurotoxic,[99,100] and PCB metabolites and congeners which exhibit diverse activities.[101-104] The potential overall contributions of individual PCB congeners to the toxicity of PCB mixtures may also involve additive or non-additive (synergism or antagonism) interactions which have not been extensively studied.[91,105,106]

[92] D. E. Schulz, G. Petrick and J. C. Duinker, *Environ. Sci. Technol.*, 1989, **23**, 852.

[93] V. A. McFarland and J. U. Clarke, *Environ. Health Perspect.*, 1989, **81**, 225.

[94] M. Mullin, G. Sawka, L. Safe, S. McCrindle and S. Safe, *J. Anal. Toxicol.*, 1981, **5**, 138.

[95] A. Parkinson, R. Cockerline and S. Safe, *Chem.-Biol. Interact.*, 1980, **29**, 277.

[96] A. Parkinson, S. Safe, L. Robertson, P. E. Thomas, D. E. Ryan and W. Levin, *J. Biol. Chem.*, 1983, **258**, 5967.

[97] A. Buchmann, W. Kunz, C. R. Wolf, F. Oesch and L. W. Robertson, *Cancer Lett.*, 1986, **32**, 243.

[98] A. Buchmann, S. Ziegler, A. Wolf, L. W. Robertson, S. K. Durham and M. Schwarz, *Toxicol. Appl. Pharmacol.*, 1991, **111**, 454.

[99] R. F. Seegal, B. Bush and W. Shain, *Toxicol. Appl. Pharmacol.*, 1990, **106**, 136.

[100] W. Shain, B. Bush and R. Seegal, *Toxicol. Appl. Pharamcol.*, 1991, **111**, 33.

[101] A. Brouwer, W. S. Blaner, A. Kukler and K. J. Van den Berg, *Chem.-Biol. Interact.*, 1988, **68**, 203.

[102] T. R. Narasimhan, H. L. Kim and S. Safe, *J. Biochem. Toxicol.*, 1991, **6**, 229.

[103] K. S. Korach, P. Sarver, K. Chae, J. A. McLachlin and J. D. McKinney, *Mol. Pharmacol.*, 1988, **33**, 120.

[104] M. C. Lans, E. Klasson-Wehler, M. Willemsen, E. Meussen, S. Safe and A. Brouwer, *Chem.-Biol. Interact.*, 1993, **88**, 7.

The adverse human impacts of PCBs have been investigated in occupationally exposed workers as well as individuals poisoned with PCB-contaminated rice oil in Japan and Taiwan (Yusho and Yu-Cheng poisonings).[91,107-109] In addition, recent studies have shown a correlation with *in utero* exposure to PCBs and subtle neurodevelopmental and neurobehavioural deficits in children.[110] These effects were observed in children with relatively low-level environmental exposure to PCBs and thus have raised concerns regarding the potential adverse effects of low level *in utero* exposure to organochlorine compounds during critical periods of foetal development.

Carcinogenesis

Several studies have reported overall cancer mortality in workers occupationally exposed to PCB mixtures and the incidence of specific cancers.[111-115] In studies containing the largest number of workers, there was not a significant overall increased incidence of cancer in these groups. However, an increased incidence of some cancers have been reported and these include melanomas, the grouping of liver, gall bladder and biliary tract cancers, gastrointestinal tract cancer in males and hematologic neoplasms. Some of the increases were not statistically significant and the carcinogenic effects observed in these workers differed in each study. These results suggest that in workers exposed to relatively high levels of PCBs there was not a consistent increase in one or more cancers; however, continued monitoring of PCB-exposed workers is warranted since laboratory animal studies in rodents have shown that some of the more highly chlorinated mixtures are hepatocarcinogens.

There has also been concern regarding the potential role of PCB, DDE and other chlorinated hydrocarbons in the development of breast cancer in women.[116-119] Three studies reported that either PCB or DDE levels were elevated in mammary tissue or serum of breast cancer patients but not in matched controls.[117-119]. However, a more recent comprehensive study of 150 breast

[105] N. Harper, K. Connor, M. Steinberg and S. Safe, *Fundam. Appl. Toxicol.*, 1995, **27**, 131.

[106] L. Biegel, M. Harris, D. Davis, R. Rosengren, L. Safe and S. Safe, *Toxicol. Appl. Pharmacol.*, 1989, **97**, 561.

[107] R. D. Kimbrough, *CRC. Crit. Rev. Toxicol.*, 1995, **25**, 133.

[108] M. Kuratsune, in *Halogenated Biphenyls, Terphenyls, Naphthalenes, Dibenzodioxins and Related Products*, ed. R. D. Kimbrough and A. A. Jensen, Elsevier, Amsterdam, 1989, p. 381.

[109] G. M. Swanson, H. E. Ratcliffe and L. J. Fischer, *Regul. Toxicol. Pharmacol.*, 1995, **21**, 136.

[110] W. J. Rogan and B. C. Gladen, *Neurotoxicology*, 1992, **13**, 27.

[111] D. P. Brown, *Arch. Environ. Health*, 1987, **42**, 333.

[112] P. A. Bertazzi, L. Ribaldi, A. Pesatori, L. Radice and C. Zocchetti, *Am. J. Ind. Med.*, 1987, **11**, 165.

[113] A. K. Bahn, I. Rosenwaike, N. Herrmann, P. Grover, J. Stellman and K. O'Leary, *New Engl. J. Med.*, 1976, **295**, 450.

[114] P. Gustavsson, C. Hogstedt and C. Rappe, *Am. J. Ind. Med.*, 1986, **10**, 341.

[115] T. Sinks, G. Steele, A. B. Smith, K. Watkins and R. A. Shults, *Am. J. Epidemiol.*, 1992, , 389.

[116] D. L. Davis, H. L. Bradlow, M. Wolff, T. Woodruff, D. G. Hoel and H. Anton-Culver, *Environ. Health Perspect.*, 1993, **101**, 372.

[117] F. Falck, A. Ricci, M. S. Wolff, J. Godbold and P. Deckers, *Arch. Environ. Health*, 1992, **47**, 143.

[118] M. S. Wolff, P. G. Toniolo, E. W. Leel, M. Rivera and N. Dubin, *J. Natl. Cancer Inst.*, 1993, **85**, 648.

[119] E. Dewailly, S. Dodin, R. Verreault, P. Ayotte, L. Sauvé, J. Morin and J. Brisson, *J. Natl. Cancer Inst.*, 1994, **86**, 232.

cancer patients in California and 150 matched controls showed that PCBs and DDE were not elevated in the former group.[120] Moreover, meta-analysis of all studies which compared organochlorine levels in breast cancer patients and controls showed that there was not a significant association between higher DDE or PCB levels and increased incidence of breast cancer.[121] It seems unlikely that background environmental exposures to these organohalogen compounds would be associated with an increased incidence of breast cancer in women since occupationally exposed workers do not have an increased incidence of this disease.[111,122]

Non-carcinogenic Responses in Occupationally Exposed Workers

Although occupational exposure to PCBs does not increase mortality, a number of PCB-related responses have been reported.[123–133] Some of the effects include chloracne and related dermal lesions; diverse hepatic effects, including increased serum levels of liver enzymes and lipids, induced hepatic drug-metabolizing enzymes and hepatomegaly; decreased birth weight of the offspring of occupationally exposed mothers; decreased pulmonary function; and eye irritation. In some studies the PCB-induced responses were reversible. It was reported that tissue levels of PCBs in transformer repair workers were higher in exposed workers than unexposed workers; however, there was no significant correlation between PCB levels and symptoms of putative PCB-induced toxicosis.[132,133] The adverse health effects of PCBs on workers exposed at toxic waste sites or during accidents or fires involving PCB-containing equipment, such as transformers or capacitors, have also been investigated.[134–137] Some neurobehavioral dysfunction may be

[120] N. Krieger, M. S. Wolff, R. A. Hiatt, M. Rivera, J. Vogelman and N. Orentreich, *J. Natl. Cancer Inst.*, 1994, **86**, 589.

[121] T. Key and G. Reeves, *Br. Med. J.*, 1994, **308**, 1520.

[122] S. Safe, *Environ. Health Perspect.*, 1995, **103**, 346.

[123] A. Fischbein, *Environ. Health Perspect.*, 1985, **60**, 145.

[124] A. Fischbein, J. N. Rizzo, S. J. Solomon and M. S. Wolff, *Br. J. Ind. Med.*, 1985, **42**, 426.

[125] A. Fischbein, J. Thornton, M. S. Wolff, J. Bernstein and I. J. Selikoff, *Arch. Environ. Health*, 1982, **37**, 69.

[126] A. Fischbein, M. S. Wolff, R. Lilis, J. Thornton and I. J. Selikoff, *Annu. N. Y. Acad. Sci.*, 1979, **320**, 703.

[127] K. Kreiss, M. M. Zack, R. D. Kimbrough, L. L. Needham, A. L. Smrek and B. T. Jones, *J. Am. Med. Assoc.*, 1981, **245**, 2505.

[128] M. Maroni, A. Columbi, G. Arbosti, S. Cantoni and V. Foa, *Br. J. Ind. Med.*, 1981, **38**, 55.

[129] R. W. Lawton, M. R. Ross, J. Feingold and J. F. Brown, *Environ. Health Perspect.*, 1985, **60**, 165.

[130] H. K. Ouw, G. R. Simpson and D. S. Siyali, *Arch. Environ. Health*, 1976, **31**, 189.

[131] R. Warshaw, A. Fischbein, J. Thornton, A. Miller and I. J. Selikoff, *Annu. N. Y. Acad. Sci.*, 1979, **320**, 277.

[132] E. A. Emmett, *Environ. Health Perspect.*, 1985, **60**, 185.

[133] E. A. Emmett, M. Maroni, J. Jefferies, J. Schmith, B. K. Levin and A. Alvares, *Am. J. Ind. Med.*, 1988, **14**, 47.

[134] P. A. Stehr-Green, V. W. Burse and E. Welty, *Arch. Environ. Health*, 1988, **43**, 420.

[135] P. W. O'Keefe and R. M. Smith, in *Halogenated Biphenyls, Terphenyls, Naphthalene, Dibenzodioxins and Related Products*, ed. R. D. Kimbrough and A. A. Jensen, Elsevier, Amsterdam, 1989, p. 417.

[136] E. F. Fitzgerald, A. L. Weinstein, L. G. Youngblood, S. J. Standfast and J. M. Melius, *Arch. Environ. Health*, 1989, **44**, 214.

[137] K. H. Kilburn, R. H. Warsaw and M. G. Shields, *Arch. Environ. Health*, 1989, **44**, 345.

associated with firemen exposed to PCBs; however, there was no correlation between levels of exposure and putative PCB-induced responses.[137]

Neurobehavioral and Neurodevelopmental Problems

Most studies which examine the high-dose effects of PCBs have focused on occupationally exposed adults. Jacobson and co-workers initiated a study to investigate the potential adverse effects of environmental contaminants on the offspring of women who consume fish from Lake Michigan.[138–141] Their results showed that there was a correlation between cord serum PCB levels and several parameters such as decreased birth weight and head circumference and neurodevelopmental deficits in newborns, which included poorer performance on the Brazelton Neonatal Behavioral Assessment Scale, on the psychomotor index of the Bayley Scales of Infant Development and on Fagan's Visual Recognition Memory Test. At four years of age there was a correlation between *in utero* exposure to PCBs and growth retardation, and reduced activity in the children was correlated with contemporary PCB body burdens. There was also a correlation between prenatal exposure to PCBs and poorer performance on the psychomotor index of the Bayley Scales of Infant Development in North Carolina infants (6 and 12 months).[142–146] The McCarthy Scales deficits reported in the Michigan study were not observed in the North Carolina children; moreover, the North Carolina children exposed to higher levels of PCBs did not exhibit unsatisfactory habits or conduct. The reason for the discrepancies between the Michigan and North Carolina studies are unknown.

A recent study in the Netherlands also investigated the potential correlation between *in utero* exposure to PCBs, PCDDs and PCDFs and neurological development in infants.[147] This study utilized high-resolution analysis of PCBs, PCDDs and PCDFs so that dioxin or toxic equivalents (TEQs) in each sample could be estimated, where

$$TEQ = \Sigma[PCB]_i \times TEF_i + \Sigma[PCDD]_i \times TEF_i + \Sigma[PCDF]_i \times TEF_i \qquad (1)$$

This method is now widely used for hazard and risk assessment of 'TCDD-like' compounds where the TEQ is the summation of the concentration of the

[138] J. L. Jacobson, S. W. Jacobson and H. E. B. Humphrey, *Neurotoxicol. Teratol.*, 1990, **12**, 319.

[139] J. L. Jacobson, S. W. Jacobson and H. E. B. Humphrey, *J. Pediatr.*, 1990, **116**, 38.

[140] S. W. Jacobson, in *PCBs: Human and Environmental Hazards*, ed. F. M. D'Itri and F. M. Kamrin, Ann Arbor Science, Ann Arbor, MI, 1983.

[141] S. W. Jacobson, G. G. Fein, J. L. Jacobson, P. M. Schwartz and J. K. Dowler, *Child Dev.*, 1985, **56**, 853.

[142] W. J. Rogan, B. C. Gladen, J. D. McKinney, N. Carreras, P. Hardy, J. Thullen, J. Tingelstad and M. Tully, *Am. J. Public Health*, 1987, **77**, 1294.

[143] B. C. Gladen, W. J. Rogan, P. Hardy, J. Thulen, J. Tingelstad and M. Tully, *J. Pediatr.*, 1988, **113**, 991.

[144] B. C. Gladen and W. J. Rogan, *J. Pediatr.*, 1991, **119**, 58.

[145] W. J. Rogan, B. C. Gladen, J. D. McKinney, N. Carreras, P. Hardy, J. Thullen, J. Tinglestad and M. Tully, *J. Pediatr.*, 1986, **109**, 335.

[146] W. J. Rogan and B. C. Gladen, *Annu. Epidemiol.*, 1991, **1**, 407.

[147] M. Huisman, C. Koopman-Esseboom, V. Fidler, M. Hadders-Algra, C. G. van der Paauw, L. G. Tuinstra, N. Weisglas-Kuperus, P. J. Sauer, B. C. Touwen and E. R. Boersma, *Early Hum. Dev.*, 1995, **41**, 111.

individual congener times its potency (TEF) relative to TCDD.[90,91,148-150] The Netherlands study utilized several measures for assessment of neurological conditions which were expressed as neurological optimality scores (NOS). There was not a significant correlation between cord or plasma PCB levels and NOS scores. However, there was a correlation between high postnatal exposure to PCBs and a higher prevalence of hypotonia. This observation contrasts with results from the North Carolina and Michigan studies, which showed correlations between neurodevelopmental deficits and prenatal exposures to PCBs. Thus, the presumed effects of *in utero* exposure to PCBs or 'dioxin-like' halogenated aromatics exhibit variability between studies, and this is an area which requires further research.

Yusho and Yu-Cheng Poisoning

Several thousand individuals were poisoned with PCBs in two separate accidents in Japan and Taiwan when PCB-containing industrial fluid accidentally leaked into rice oil which was then purchased by consumers.[108,151-155] The toxic symptoms observed in both the Yusho (Japan) and Yu-Cheng (Taiwan) poisonings included severe and persistent chloracne, dark brown pigmentation of nails, distinctive hair follicles, skin thickening, various ocular problems, numbness in some extremities and numerous subjective complaints that could be associated with neurological problems. The severity of these acute poisonings with PCBs was somewhat surprising since most of these symptoms are not observed in individuals occupationally exposed to PCBs. Moreover, adipose tissue and serum analysis of Yusho and Yu-Cheng victims, workers and normal individuals showed that their PCB levels were comparable. In contrast, the corresponding PCDF concentrations were consistently higher in the Yusho and Yu-Cheng patients. Subsequent fractionation of the contaminated oils and laboratory animal studies have demonstrated that the PCDFs present in these mixtures were significantly more toxic than the PCBs in the Yushu/Yu-Cheng oil.[156-158]

[148] U. G. Ahlborg, A. Brouwer, M. A. Fingerhut, J. L. Jacobson, S. W. Jacobson, S. W. Kennedy, A. A. F. Kettrup, J. H. Koeman, H. Poiger, C. Rappe, S. H. Safe, R. F. Seegal, J. Tuomisto and M. Van den Berg, *Eur. J. Pharmacol.*, 1992, **228**, 179.

[149] U. G. Ahlborg, A. Hanberg and K. Kenne, *Risk Assessment of Polychlorinated Biphenyls (PCBs)*, Institute of Environmental Medicine, Karolinska Institute, Stockholm, Sweden, 1992.

[150] U. G. Ahlborg, G. C. Becking, L. S. Birnbaum, A. Brouwer, H. J. G. M. Derks, M. Feeley, G. Golor, A. Hanberg, J. C. Larsen, A. K. D. Liem, S. Safe, C. Schlatter, F. Arn, M. Younes and E. Yrjänheikki, *Chemosphere*, 1994, **28**, 1049.

[151] M. Kuratsune, *Environ. Health Perspect.*, 1972, **1**, 119.

[152] Y. C. Lu and P. N. Wong, *Am. J. Ind. Med.*, 1984, **5**, 81.

[153] Y. C. Lu and Y. C. Wu, *Environ. Health Perspect.*, 1985, **59**, 17.

[154] M. Kuratsune and R. E. Shapiro, *PCB Poisoning in Japan and Taiwan*, Liss, New York, 1984.

[155] W. J. Rogan, in *Halogenated Biphenyls, Terphenyls, Naphthalenes, Dibenzodioxins and Related Products*, ed. R. D. Kimbrough and A. A. Jensen, Elsevier, Amsterdam, 1989, p. 401.

[156] T. Kashimoto, H. Miyata, S. Kunita, T. C. Tung, S. T. Hsu, K. J. Chang, S. Y. Tang, G. Ohi, J. Nakagawa and S. I. Yamamoto, *Arch. Environ. Health*, 1981, **32**, 321.

[157] Y. Masuda, H. Kuroki, K. Haraguchi and J. Nagayama, *Environ. Health Perspect.*, 1985, **59**, 53.

[158] N. Kunita, T. Kashimoto, H. Miyata, S. Fukushima, S. Hori and H. Obana, *Am. J. Ind. Med.*, 1984, **5**, 45.

The effects of *in utero* exposure to the contaminated rice oil resulted in a host of toxic effects in the Yu-Cheng infants, which included pigmented nails, fingers and toes, gum hypertrophy, tooth chipping, acne, hyperpigmentation, conjunctivitis or cysts and hirsutism.[159] In addition, the effects of *in utero* exposure to the PCBs/PCDFs from the rice oil on cognitive development and behaviour in the Yu-Cheng children were determined using Rutter's Child Behavior Scale, the Bayley Scale for Infant Development, the Stanford–Binet Test and the Wechsler Intelligence Scale for Children.[160–163] In all of these tests the PCB/PCDF-exposed children exhibited some deficits, although the deficits were not significant at every time-point. Comparison between the deficits in Yusho/Yu-Cheng, North Carolina and Michigan children has not been carried out, and it is not possible to delineate which responses are Ah receptor-dependent or -independent. The early developmental effects of PCBs, PCDDs and PCDFs and the possible contribution of 'natural' AhR agonists is an area which requires further research.

5 Halogenated Aromatics and the Endocrine Disruption Hypothesis

In recent years there has been growing concern by some scientists that persistent organochlorine compounds may be responsible for the hypothesized decrease in male sperm counts, increase in male reproductive problems and breast cancer in women.[116,164,165] These hypotheses are based on laboratory animal studies and observations in fish and wildlife populations. For example, halogenated aromatics such as PCBs, PCDDs and PCDFs as well as DDT, DDE and related compounds can act as hormone mimics, particularly estrogens, and disrupt various endocrine response pathways. Moreover, *p,p'*-DDE, a major environmental contaminant, has recently been characterized as an antiandrogen in rodents and mammalian cells in culture.[166] It has also been pointed out that there are a host of naturally occurring compounds in foods or cooked foods which bind to the AhR (*e.g.* indole-3-carbinol, aromatic amines, polynuclear aromatic hydrocarbons), estrogen receptor (bioflavonoids, lignans), retinoic acid receptor (vitamin and retinoid precursor) and other cellular receptors.[122] The relative contributions (mass and potency) of natural components in raw and cooked foods must be compared to the relative contributions of organic pollutants for various endocrine response pathways. It has recently been suggested that the estrogen and antiestrogen equivalents of natural endocrine disrupters on food far outweigh contributions from xenoestrogens.[122] These issues must be addressed

[159] W. J. Rogan, B. C. Gladen, K. Hung, S. Koong, L. Shih, J. S. Taylor, Y. Wu, D. Yang, N. B. Ragan and C. Hsu, *Science*, 1988, **241**, 334.

[160] Y.-C. J. Chen, Y.-L. Guo, C.-C. Hsu and W. J. Rogan, *J. Am. Med. Assoc.*, 1992, **268**, 3213.

[161] T.-J. Lai, Y.-L. Guo, M.-L. Yu, H.-C. Ko and C.-C. Hsu, *Chemosphere*, 1994, **29**, 2405.

[162] M.-L. Yu, C.-C. Hsu, Y.-L. Guo, T.-J. Lai, S-J. Chen and J.-M. Luo, *Chemosphere*, 1994, **29**, 2413.

[163] Y.-L. Guo, T.-J. Lai, S-J. Chen and C.-C. Hsu, *Bull. Environ. Contam. Toxicol.*, 1995, **55**, 8.

[164] R. M. Sharpe and N. F. Skakkebaek, *Lancet*, 1993, **341**, 1392.

[165] T. Colborn, F. S. Vom Saal and A. M. Soto, *Environ. Health Perspect.*, 1993, **101**, 378.

[166] W. R. Keice, C. R. Stone, S. C. Laws and L. E. Gray, *Nature*, 1995, **375**, 581.

in future studies on hazard and risk assessment of organochlorine pollutants as endocrine disrupters.

6   Acknowledgements

The financial assistance of the National Institutes of Health (ESO4917) is gratefully acknowledged.

Ecotoxicology of Chlorinated Aromatic Hydrocarbons

PIM DE VOOGT

1 Introduction

One of the consequences of the growth of the human population, the corresponding increasing need for food and health, and the economic necessity of an ever extending market, has been the dramatic increase in the production of man-made chemicals. Among these the halogenated organic compounds, which have been relatively easy as well as cheap to manufacture, have attracted much attention because of their large production volumes and persistent character, resulting in the discovery of their presence in the environment. In particular, the chlorinated aromatic compounds (CACs) and chlorinated pesticides are important in this respect, since several biological effects observed in avian and mammalian populations in the field have been associated with the environmental contamination levels of these compounds.[1]

The ecotoxicologically important CACs can be typically divided into compounds that have been manufactured commercially and those that are formed unintentionally or released incidentally in production processes. Examples of the former group are the chlorinated benzenes, including the pesticides hexachlorobenzene, dichlorodiphenyltrichloroethane(p,p'-DDT) and the technical products polychlorinated naphthalenes(PCNs),[2] the polychlorinated biphenyls(PCBs) and their substitutes in industrial applications, *viz.* the tetrachlorobenzyltoluenes(TCBTs).[3] The most important compounds belonging to the latter group are the polychlorinated dibenzo-*p*-dioxins (PCDDs), dibenzofurans (PCDFs) and related compounds, including the chlorinated diphenyl ethers[4] and dibenzothiophenes. A third category is the group of CACs of natural, non-man-made origin, of which still relatively little is known[5,6] and which will not be dealt with in this review.

[1] K. J. Canters and G. R. de Snoo, *Rev. Environ. Contam. Toxicol.*, 1992, **130**, 1.
[2] U. Järnberg, L. Asplund, C. de Wit, A. K. Grafström, P. Haglund, B. Jansson, K. Lexén, M. Strandell, M. Olsson and B. Jonsson, *Environ. Sci. Technol.*, 1993, **27**, 1364.
[3] A. G. van Haelst, Q. Zhao, F. van der Wielen, H. A. J. Govers and P. de Voogt, *Ecotoxicol. Environ. Safe.*, 1996 in press.
[4] M. Becker, T. Phillips and S. Safe, *Toxicol. Environ. Chem.*, 1991, **33**, 189.
[5] W. Fenical, in *Marine Organic Chemistry*, ed. E. K. Duursma and R. Dawson, Elsevier, Amsterdam, 1981, ch. 12.
[6] E. de Jong, J. A. Field, H. E. Spinnler, J. B. P. A. Wijnbergand and J. A. M. de Bont, *Appl. Environ. Microbiol.*, 1994, **60**, 264.

The environmental chemistry, including the behaviour and environmental fate, of these compounds has been reviewed extensively.[7-11] Their emissions to the environment are almost exclusively by way of commercial, technical mixtures[12] or mixtures related to certain formation processes, such as, for example, waste incineration or production of chlorinated organic products (*e.g.* pentachlorophenol) and intermediates (*e.g.* chlorinated benzenes). Such mixtures may contain many hundreds of congeners (see Table 1). Hence, residues found in the environment are also a complex mixture of congeners and isomers. The most striking phenomenon in environmental residues, however, is that their congener profiles are highly variable and no longer resemble the original technical mixtures, due to different distribution properties of the congeners and abiotic and biotic transformations.[13]

Most of the CACs mentioned have a global dispersion nowadays. This is probably a result of the relatively large fluxes from water and land to the atmosphere of pesticides and organochlorines in the areas of application/use and their relatively long lifetimes in the environment, whereafter redistribution in colder areas such as the temperate zones and arctic environments will occur.[14-16] Since many industrialized countries banned or reduced the production and/or use of chlorinated industrial mixtures, such as PCBs, and pesticides, *e.g.* DDT and drins, in the 1970s and 1980s, recent sources of global contamination by organochlorine residues are most likely emissions in the tropical belt.[14] Through long-range atmospheric transport[16] the persistent compounds are conveyed to remote, clean areas and it is now believed that oceans, in particular Arctic waters, have become a sink.[17] The emissions of PCDDs and PCDFs have been reduced as a result of technical measures taken in waste incineration and production processes, but for obvious reasons such measures have been taken primarily in developed countries.

In the literature, the wildlife ecotoxicology of CACs has not been dealt with to the extent of agrochemicals.[18] The purpose of this chapter is to review eco-toxicological aspects of chlorinated aromatic compounds, with emphasis on polychlorinated biphenyls, dibenzo-*p*-dioxins and dibenzofurans. Rather than being complete we will try to review recent information on bioaccumulation, biotransformation and effects of CACs. The focus will be on avian and mammalian

[7] *PCBs and the Environment*, ed. J. S. Waid, CRC Press, Boca Raton, 1986.

[8] P. de Voogt, D. E. Wells, L. Reutergårdh and U. A. T. Brinkman, *Int. J. Environ. Anal. Chem.*, 1990, **40**, 1.

[9] V. Lang, *J. Chromatogr.*, 1992, **595**, 1.

[10] K. Ballschmiter, C. Rappe and H. R. Buser, in *Halogenated Biphenyls, Terphenyls, Naphthalenes, Dibenzodioxins and Related Products*, ed. R. D. Kimbrough and A. A. Jensen, Elsevier, Amsterdam, 1989, ch. 2.

[11] C. Rappe and H. R. Buser, in *Halogenated Biphenyls, Terphenyls, Naphthalenes, Dibenzodioxins and Related Products*, ed. R. D. Kimbrough and A. A. Jensen, Elsevier, Amsterdam, 1989, ch. 3.

[12] P. de Voogt and U. A. T. Brinkman, in *Halogenated Biphenyls, Terphenyls, Naphthalenes, Dibenzodioxins and Related Products*, ed. R. D. Kimbrough and A. A. Jensen, Elsevier, Amsterdam, 1989, ch. 1.

[13] S. H. Safe, *Crit. Rev. Toxicol.*, 1994, **24**, 87.

[14] S. Tanabe, H. Iwata and R. Tatsukawa, *Sci. Total Environ.*, 1994, **154**, 163.

[15] H. Iwata, S. Tanabe, N. Sakai, A. Nishimura and R. Tatsukawa, *Environ. Pollut.*, 1994, **85**, 15.

[16] P. de Voogt and B. Jansson, *Rev. Environ. Contam. Toxicol.*, 1993, **132**, 1.

[17] H. Iwata, S. Tanabe, S. Sakai and R. Tatsukawa, *Environ. Sci. Technol.*, 1993, **27**, 1080.

[18] R. J. Kendall and J. Akerman, *Environ. Toxicol. Chem.*, 1992, **11**, 1727.

Table 1 Multiplicity of halogenated aromatic compounds: number of isomers and congeners[a]

No. of Cl atoms	1	2	3	4	5	6	7	8	9	10	Total
PCB/ PBB/ PCDE	3	12	24	42	46	42	24	12	3	1	209
PCDF/ PBDF	4	16	28	38	28	16	4	1	–	–	135
PCDD/ PBDD/ PCN	2	10	14	22	14	10	2	1	–	–	75
R$_1$PCB[b]	19	64	136	198	198	136	64	19	3	–	837
DDTs				4	2						6
TCBT				96							96
PCC											13824
PCT											8149
PCBDD											5020

[a]PCB, polychlorinated biphenyl; PBB, polybrominated biphenyl; PCDE, polychlorinated diphenyl ether; PCDF, polychlorinated dibenzofurans; PCDD, polychlorinated dibenzo-*p*-dioxins; PCN, polychlorinated naphthalenes; PCBDD, polychlorobromodibenzo-*p*-dioxins; PCC, polychlorinated camphenes (toxaphene); PCT, polychlorinated terphenyls; DDTs, *p,p'*-DDT, its metabolites DDE and DDD and impurities (*o,p*-isomers); TCBT, tetrachlorobenzyltoluene (dichlorobenzyldichlorotoluene, ugilec).
[b]R = *e.g.* CH_3, OH, CH_3SO_2.

wildlife species, since most of the literature until now is centred around vertebrates and because these species are considered to be more at risk from the compounds considered here than invertebrates.

2 Bioaccumulation of CACs

For gill-breathing animals, both the surrounding water and the food contribute to the total body burden of CACs. Quite often a strong correlation is observed in these animals between water solubilities of the various (congeners of) CACs and bioconcentration factors.[19] This indicates that gill uptake is probably the major source and that equilibrium partitioning determines their degree of bioaccumulation. For birds and marine mammals, food is the main source for organochlorines, and metabolism and elimination (*e.g.* through lactation) primarily determine the actual body burden.[20,19]

Background levels of PCBs are found in aquatic mammals from areas all over

[19] S. Tanabe and R. Tatsukawa, in *Persistent Pollutants in Marine Ecosystems*, eds. C. H. Walker and D. R. Livingstone, Pergamon Press, Oxford, 1992, ch. 7.
[20] J. P. Boon, E. van Arnhem, S. Jansen, N. Kannan, G. Petrick, D. Schulz, J. C. Duinker, P. J. H. Reijnders and A. Goksøyr, in *Persistent Pollutants in Marine Ecosystems*, ed. C. H. Walker and D. R. Livingstone, Pergamon Press, Oxford, 1992, ch. 6.

the world. The highest levels of PCBs and DDTs have been detected in coastal species.[19] Lowest values are found in the southern hemisphere, with levels 1000 times less than those found in some areas in the northern hemisphere, notably the North and Baltic Seas.[19,21] This geographical distribution can be explained by the former location of production sites and extensive use in developed countries. As pointed out, dispersion is more global nowadays and geographical trends may level off and reflect more the redistribution phenomenon discussed above.

Time trend analyses in uncontaminated[22,23] and contaminated areas[24-26] have shown that PCB and DDT levels in marine and freshwater organisms increased steadily until the mid-1970s, following the variation in industrial production of both products with a 5–10 years lag phase. After 1974–77, PCB levels tend to decrease to a pre-1970 level[23] and then level-off in the 1980s, whereas for DDTs a still continuing decrease is found in the last decade. For hexachlorocyclohexanes (HCHs), on the contrary, levels monitored in seals from 1970 to 1988 do not show a clear increasing or decreasing trend,[23] probably as a result of ongoing use in the tropics and subsequent long-range transport.[14]

From the total of 210 PCDD and PCDF congeners, only the 2,3,7,8-substituted congeners are recalcitrant. Although non-2,3,7,8-substituted congeners can be found to accumulate in lower organisms, *e.g.* crustaceans, in general, the remaining congeners are transformed in mammals, birds and fish. As a consequence, isomeric patterns in biological samples tend to be similar.[27] Differences can be observed between aquatic and terrestrial organisms, however. Concentrations of PCDDs and PCDFs in terrestrial animals are usually much lower than those in aquatic species, even if top predators are concerned.[26]

Time trend analyses in guillemots from Scandinavia have revealed steadily decreasing PCDD/F concentrations. In the period from 1969 to 1992, concentrations dropped to 70–80% of the initial levels.[26] This may reflect the fact that PCB mixtures have been the most important sources of PCDD and PCDF contamination until their ban in the 1970s, at least in certain parts of the world.

Body Distribution

Laboratory experiments have shown that vertebrate body distribution of CACs is governed by their affinity for specific binding sites in the liver relative to their affinity for non-specific binding sites in liver and extrahepatic tissues and the

[21] M. A. Kamrin and R. K. Ringer, *Toxicol. Environ. Chem.*, 1994, **41**, 63.

[22] S. Tanabe, J.-K. Sung, D.-Y. Choi, N. Baba, M. Kiyota, K. Yoshida and R. Tatsukawa, *Environ. Pollut.*, 1994, **85**, 305.

[23] L. M. Hernández, M. C. Rico, M. J. González, M. Hernan and M. A. Fernández, *J. Field Ornithol.*, 1986, **57**, 270.

[24] M. Olsson and L. Reutergårdh, *Ambio*, 1986, **15**, 103.

[25] R. J. Fensterheim, *Regul. Toxicol. Pharmacol.*, 1993, **18**, 181.

[26] C. de Wit, U. A. Järnberg, T. Asplund, B. Jansson, M. Olsson, T. Odsjö, I. L. Lindstedt, Ö. Andersson, S. Bergek, M. Hjelt, C. Rappe, A. Jansson and M. Nygren, *Organohalogen Compd.*, 1994, **20**, 47.

[27] C. Rappe, R. Andersson, P.-A. Bergqvist, C. Brohede, M. Hansson, L. -O. Kjeller, G. Lundström, S. Marklund, M. Nygren, S. E. Swanson, M. Tysklind and K. Wiberg, *Chemosphere*, 1987, **16**, 1603.

number of available binding sites.[28] Non-specific sites include lipids, present in most body tissues, possessing high capacities.[29] Specific sites are the aryl hydrocarbon (Ah) receptor (present in liver and other tissues, *e.g.* kidney) and the CYP1A2 protein (present in liver).[29] The affinities, in turn, are governed by physicochemical properties of the compounds. Binding sites can be divided into inducible and non-inducible sites. The number present of the former depends on the amount of CACs bound to the Ah receptor. Here, non-additive (antagonistic) effects can be important, in particular when the organism is exposed to complex mixtures, as is the case in wildlife.

Liver and adipose tissue are the major storage compartments for CACs. Ratios between liver and adipose tissue distribution are species dependent and follow the order rodents > birds > humans > fish.[30] In marine mammals this ratio is determined by their relatively large adipose tissue compartment, with large storage capacities. CAC metabolites may bind to specific target proteins, *e.g.* in blood, kidney or lung.

Mammals

Because oceans serve as a sink for chlorinated pesticides and CACs, marine mammals and piscivorous birds are consuming relatively high amounts of these compounds, whereas they have a relatively low capacity for metabolism. As a result, they may constitute the most vulnerable organisms with regard to long-term toxicity.[14] Over 90% of the whole body burden of marine mammals may be present in their blubber.[14]

Adult male pinnipeds show increasing levels of PCBs and DDTs with age, whereas females appear to diminish their body burden through transplacental transfer and, more importantly, weaning of their kids.[14,31] The same phenomenon has been observed in cetaceans[32] and turtles.[33] After the menopause, levels may slightly increase again in females.[23] Terrestrial mammals may also secrete CACs through their anal glands.[34]

Separately living populations of wildlife species may show differences in patterns of environmental contaminants like PCBs. PCB and PCDD/F congener patterns may differ in the same species from various locations as a result of the difference in patterns already present in their (differing) food, as has been demonstrated for seals fed with fish.[35] Such differences also relate to selective biotransformation capacities of the organisms involved.[20] Large differences in metabolic capacity have been observed between pinnipeds and cetaceans (*cf.* biotransformation section). This does not only hold for CACs, but also for chlorinated pesticides. For instance, porpoises have been shown to have a

[28] J. de Jongh, Ph.D. Thesis, University of Utrecht, 1994.
[29] M. E. Andersen, J. J. Mills, M. L. Gargas, L. B. Kedderis, L. S. Birnbaum, D. Neubert and W. F. Greenlee, *Risk Anal.*, 1993, **13**, 25.
[30] M. van den Berg, J. de Jongh, H. Poiger and J. R. Olson, *Crit. Rev. Toxicol.*, 1994, **24**, 1.
[31] A. Abarnou, D. Robineau and P. Michel, *Oceanol. Acta*, 1986, **9**, 19.
[32] A. Borrell, *Mar. Pollut. Bull.*, 1993, **26**, 146.
[33] L. Meyers-Schöne and B. T. Walton, *Rev. Environ. Contam. Toxicol.*, 1994, **135**, 93.
[34] P. E. G. Leonards, B. van Hattum, W. P. Cofino and U. A. T. Brinkman, *Environ. Toxicol. Chem.*, 1994, **13**, 129.
[35] E. Storr-Hansen, H. Spliid and J. P. Boon, *Arch. Environ. Contam. Toxicol.*, 1995, **28**, 48.

relatively low potential for metabolism of PCBs, DDT and HCH compared to several pinnipeds from the same area.[23]

Data on PCDD and PCDF concentrations in wildlife are scarce and mainly available for aquatic species. OctaCDD, 2,3,7,8-tetraCDD and 1,2,3,7,8-penta-CDD are the most abundant PCDD congeners in marine mammals and 2,3,7,8-tetraCDF and 2,3,4,7,8-pentaCDF the most abundant PCDFs.[36] Hexa- and heptachlorinated dioxin and furan concentrations are usually very low in marine mammals and may point to relatively high metabolic and elimination rates.[37]

The decline of otter populations in large parts of Europe has been suggested to be causally linked to contamination with organochlorines and in particular to PCBs. This is because average PCB levels observed in otter are highest in populations in decline[38] and well above levels at which a closely related mustelid species, viz. the American mink, which like otter feeds mainly on fish and other aquatic fauna, elicits reproductive failure in laboratory studies.[39,40] PCDD and PCDF patterns in otter show that hepta- and octachlorinated congeners are more abundant than tetra-, penta- (except for 2,3,4,7,8-pentaCDF) or hexa-chlorinated congeners,[41] contrary to what has been found in marine mammals.

Even less is known about CAC levels in terrestrial mammals and only a few studies on congener specific PCBs, PCDDs and PCDFs in such organisms are available. Scandinavian studies have shown that levels of PCBs and PCDD/Fs in fox, lynx, reindeer, badger and moose are generally quite low when compared to purely aquatic species.[26,42] In particular, fox appears to have a markedly high metabolic capacity for PCBs.[42] PCB patterns observed in various mustelids indicate that these also have high metabolizing capacities[34,43] and that corresponding PCB metabolites can be found in adipose tissue and liver.[44] PCDD and PCDF concentrations in these animals contribute to about half of the toxic equivalent concentration (see below), the other half being contributed by the planar PCBs.[41] Pattern comparison is less frequently done with PCDD and PCDF congener profiles, because selection of a reference compound is less unambiguous than with PCBs (see below), due to possible breakdown of all congeners. Having said this, comparison of patterns may still elucidate some relevant (dis)similarities between metabolic potency of species. The observed congener pattern for PCDDs and PCDFs in terrestrial mustelids and also in otter is different from that observed in marine mammals[36] (see Figures 1 and 2) and resemble the patterns found in human fat and milk and in experimental studies

[36] H. P. Goorissen, Comparison of PCDD and PCDF Congener Patterns in Different Species and Trophic Levels: Evidence for Metabolism?, University of Amsterdam, Amsterdam, 1994 (in Dutch).
[37] S. Bergek, P.-A. Bergqvist, M. Hjelt, M. Olsson, C. Rappe, A. Roos and D. Zook, Ambio, 1992, 21, 553.
[38] M. D. Smit, P. E. G. Leonards, B. van Hattum and A. W. J. J. de Jongh, PCBs in European Otter (Lutra lutra) Populations, Vrije Universiteit, Amsterdam, 1994.
[39] J. E. Kihlström, M. Olsson, S. Jensen, Å. Johansson, J. Ahlbom and Å. Bergman, Ambio, 1992, 21, 563.
[40] P. E. G. Leonards, T. H. de Vries, W. Minnaard, S. Stuijfzand, P. de Voogt, W. P. Cofino, N. M. van Straalen and B. van Hattum, Environ. Toxicol. Chem., 1995, 14, 639.
[41] P. de Voogt, M. J. M. van Velzen and P. E. G. Leonards, Organohalogen Compd., 1994, 21, 465.
[42] L. Asplund, A. Olsson, L. Häggberg, M. Athanasiadou, M. Olsson and Å. Bergman, Abstracts, Annual Meeting SETAC Europe, Copenhagen, 1995, p. 210.
[43] P. E. G. Leonards, S. Broekhuizen, B. van Hattum, P. de Voogt, U. A. T. Brinkman, N. M. van Straalen and W. P. Cofino, Organohalogen Compd., 1993, 14, 101.
[44] P. de Voogt, M. J. M. van Velzen and P. E. G. Leonards, Organohalogen Compd., 1993, 14, 105.

Figure 1 Relative abundance of 2,3,7,8-substituted PCDD congeners to octachlorodibenzo-*p*-dioxin in marine (a) and terrestrial mammals (b). Each congener is indicated with its chlorine substitution pattern in addition to the 2,3,7,8-substitution and its total number of chlorines, *i.e.* 1P5 means 1,2,3,7,8-pentaCDD. When no bars are given, the corresponding congener was not determined. When a congener was below the limit of detection (lod), it was assigned the lod value, or, if lod not given, the value of 0.001. H's dolphin = Hector's dolphin

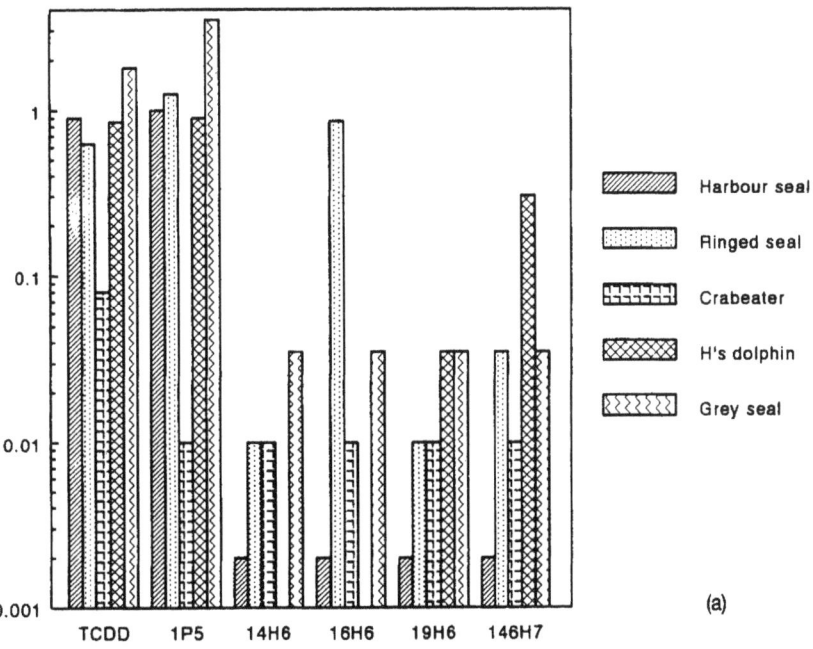

(a)

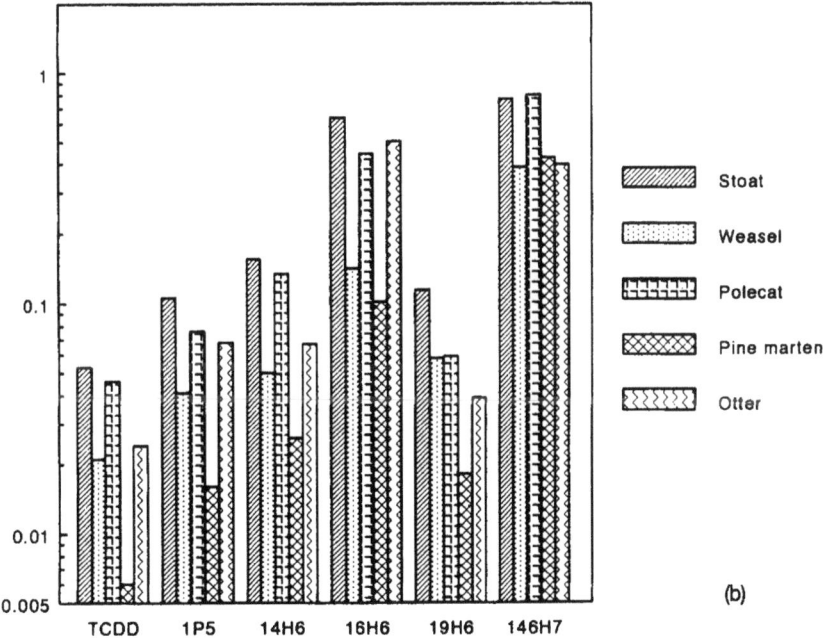

(b)

Figure 2 Relative abundance of 2,3,7,8-substituted PCDF congeners to 2,3,4,7,8-pentaCDF in marine (a) and terrestrial mammals (b). For explanation, see legend to Figure 1

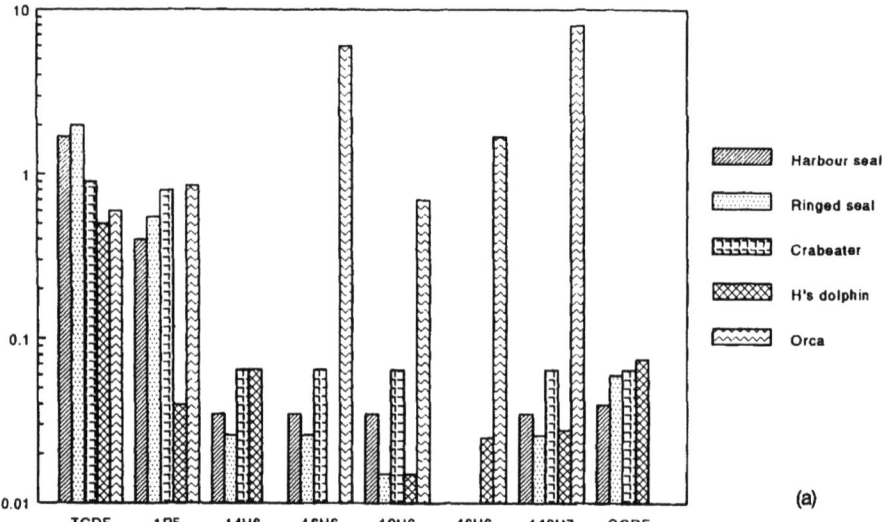

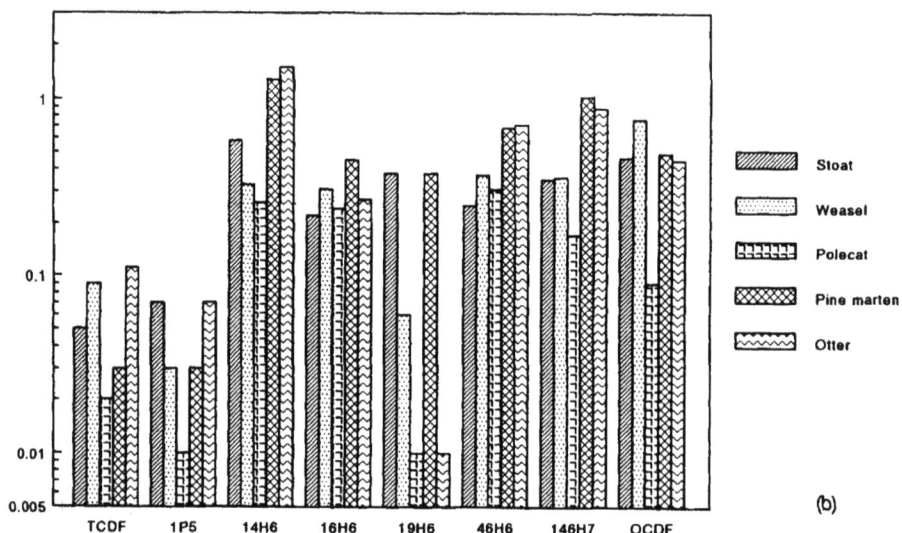

with hamster.[30] In mustelids from North America, PCB and chlorinated insecticide levels declined in a ten year period (1972–1981).[45]

Birds

Concentrations of PCBs, PCDDs and PCDFs in eggs of fish-eating birds have been reviewed recently.[46] Congener analysis in birds from North America and Europe showed that sources of PCBs may be different in these continents.[46] PCDDs and PCDFs contribute only to a minor extent to the total dioxin toxicity observed in bird eggs; the majority is caused by non- and mono-*ortho*-substituted PCB congeners.

Species differences in CAC concentrations and congener patterns can also be observed in birds. Marine or fish-eating birds generally show higher concentrations of PCBs than terrestrial birds of prey.[47,48] This is due to slow metabolic rates in birds with specified diets as a result of low levels of cytochrome P450 isozymes[47] (see below). Congener patterns of PCDDs and PCDFs (Figure 3) show that the abundance of congeners relative to the reference compound (OCDD and 2,3,4,7,8-pentaCDF, respectively) tend to be somewhat higher than in marine mammals. In small terrestrial birds a pattern similar to that of terrestrial mammals is observed (*i.e.* a higher abundance of the higher chlorinated congeners), whereas fish-eating birds show a pattern more similar to that of marine mammals (*cf.* cormorant) or intermediate to terrestrial and marine species (common tern).

Adult birds have significantly higher concentrations of organochlorines than juveniles, although sex may interfere with this trend (males displaying higher concentrations than females), probably as a result of transfer to eggs in females.[49] Birds species feeding on comparable food sources show less differences in congener patterns. Intraspecies comparison between waders from the Wadden Sea showed differences in absolute PCB concentrations but similar congener patterns.[49] Migratory birds like waders may show distinct concentration increases in organochlorine concentrations during moulting and when residing temporarily in contaminated areas.[50]

Time trend analysis in bird eggs has shown that levels of PCBs and DDTs have decreased from the 1970s until now, on both sides of the Atlantic Ocean.[23,51,52]

[45] T. Sleeves, M. Strickland, R. Frank, J. Rasper and C. W. Douglas, *Bull Environ. Contam. Toxicol.,* 1991, **46**, 368.

[46] A. T. C. Bosveld and M. van den Berg, *Environ. Rev.,* 1994, **2**, 147.

[47] C. H. Walker, in *Persistent Pollutants in Marine Ecosystems,* ed. C. H. Walker and D. R. Livingstone, Pergamon Press, Oxford, 1992, ch. 10.

[48] R. S. Boumphrey, S. J. Harrad, K. C. Jones and D. Osborn, *Arch. Environ. Contam. Toxicol.,* 1993, **25**, 346.

[49] J. M. Everaarts, A. de Buck, M. T. J. Hillebrand and J. P. Boon, *Sci. Total Environ.,* 1991, **100**, 483.

[50] P. de Voogt, J. C. Klamer, A. A. Goede and H. Govers, *Accumulation of Organochlorine Compounds in Waders from the Dutch Wadden Sea,* Vrije Universiteit, Amsterdam, IvM report 85/7, 1985.

[51] P. A. Pearce, J. E. Elliott, D. B. Peakall and R. J. Norstrom, *Environ. Pollut.,* 1989, **56**, 217.

[52] D. G. Noble and S. P. Burns, *State of Environment Fact Sheet 90-1,* Environment Canada, Ottawa, 1990, p. 1; see also ref. 106.

Figure 3 Relative abundance of 2,3,7,8-substituted PCDD congeners to octachlorodibenzo-*p*-dioxin (a) and 2,3,7,8-substituted PCDF congeners to 2,3,4,7,8-pentaCDF (b) in eggs of some aquatic and terrestrial birds. Rw. blackbird = red-winged blackbird, Gr. bl. heron = great blue heron, Dc. cormorant = double-crested cormorant. For further explanation, refer to Figure 1

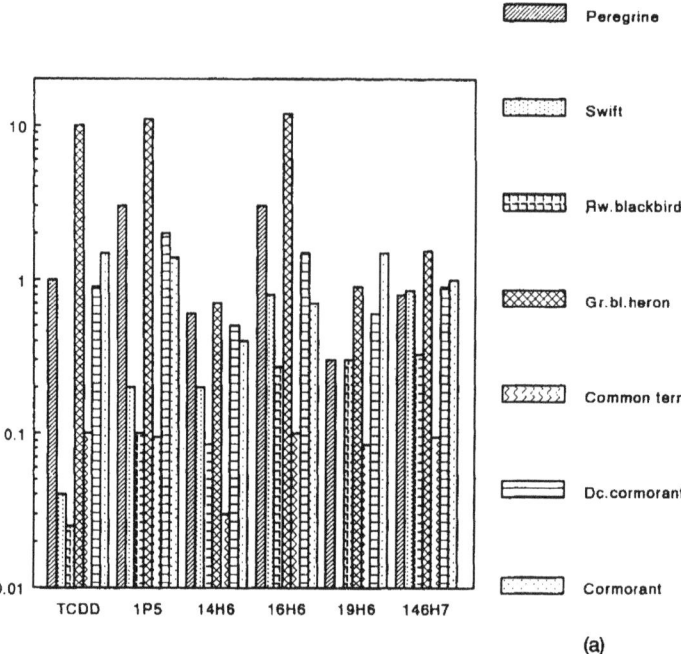

(a)

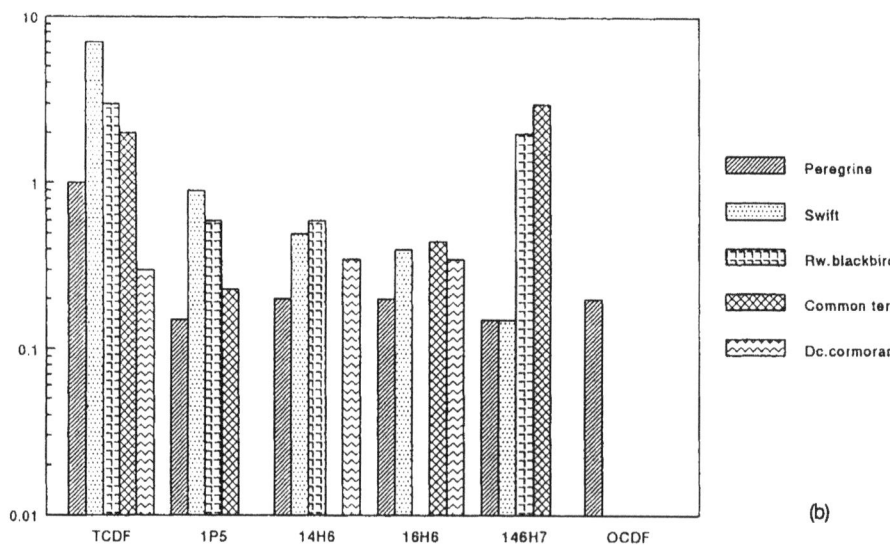

(b)

3 Biotransformation

Transformation of CACs by organisms can be divided into biodegradation and biotransformation. Biodegradation usually refers to transformation by micro-organisms, and includes both oxidative and reductive degradation. Oxidative degradation in micro-organisms is mediated through dioxygenases, contrary to higher organisms where monooxygenases are involved in the transformation of CACs.

Relatively little is known about the biotransformation of CACs in lower phyla in the environment.[53] In aquatic fauna, including molluscs, annelids, echinoderms and larger crustaceans, alterations of the CYP1A type (*cf.* below) seem to be absent, while present in fish. Larger crustaceans and fish can probably transform globular PCBs, whereas molluscs, annelids and echinoderms cannot.[54] However, there is evidence for monooxygenase activities in marine invertebrates.[55] Because insufficient information is available to draw general conclusions, here attention will be focused on biotransformation by higher vertebrates. The interested reader is referred to two recent reviews on invertebrates.[53,55] In the following, a brief treatise on biodegradation will be presented first. Subsequently, biotransformation mechanisms of CACs will be discussed in more detail, followed by an overview of metabolites patterns and levels found in wildlife. Finally, biotransformation of some chlorinated pesticides will be addressed.

Biodegradation

Halogenated aromatic compounds can be degraded by many types of micro-organisms. The removal of halogen substituents is a key step in biodegradation.[56] Dehalogenation can occur through three classes of reactions, *viz.* oxidative, hydrolytic or reductive dehalogenation, respectively. Oxidative dehalogenation occurs under aerobic conditions and involves fortuitous loss of halogen substituents during oxygenation of the aromatic ring. Examples are the degradation of chlorinated benzoate herbicides (*e.g.* dicamba, 3,6-dichloro-2-methoxybenzoate) and chlorinated benzenes to catechols (*cf.* Scheme 1).

Hydrolytic dehalogenation occurs in both aerobic and denitrifying conditions and involves replacement of the halogen by a hydroxy group with water as the source of the oxygen atom.[56] Reductive chlorination occurs almost exclusively under anaerobic conditions and involves subsequent substitution of halogen atoms by hydrogen. Chlorinated benzenes and phenols such as hexachlorobenzene and pentachlorophenol can be partly or even totally dehalogenated by reductive dehalogenation.

Aerobic biodegradation of PCBs has been studied extensively. The following generic relationships have been derived from these studies: (i) less chlorinated PCBs are more quickly degraded, (ii) dioxygenation occurs at the least-substituted ring, (iii) PCBs with substituents on both rings are more recalcitrant than congeners containing an unsubstituted ring. Aerobic degradation of PCBs usually involves 2,3-dioxygenation and subsequent ring cleavage resulting in chlorinated benzoates (see Scheme 2).

[53] W. Ernst, in *Persistent Pollutants in Marine Ecosystems*, ed. C. H. Walker and D. R. Livingstone, Pergamon Press, Oxford, 1992, ch. 2.

[54] J. F. Brown, Jr., *Mar. Environ. Res.*, 1992, **34**, 261.

[55] D. R. Livingstone, in *Persistent Pollutants in Marine Ecosystems*, ed. C. H. Walker and D. R. Livingstone, Pergamon Press, Oxford, 1992, ch. 1.

[56] L. C. M. Commandeur and J. R. Parsons, in *Biochemistry of Microbial Degradation*, ed. C. Ratledge, Kluwer, Dordrecht, 1993, p. 423.

P. de Voogt

Scheme 1 Oxidative dehalogenation of chlorinated benzenes

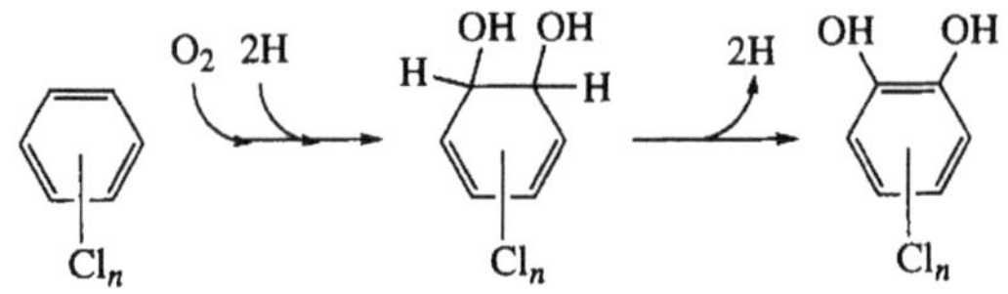

Scheme 2 Pathways for the aerobic biodegradation of PCBs

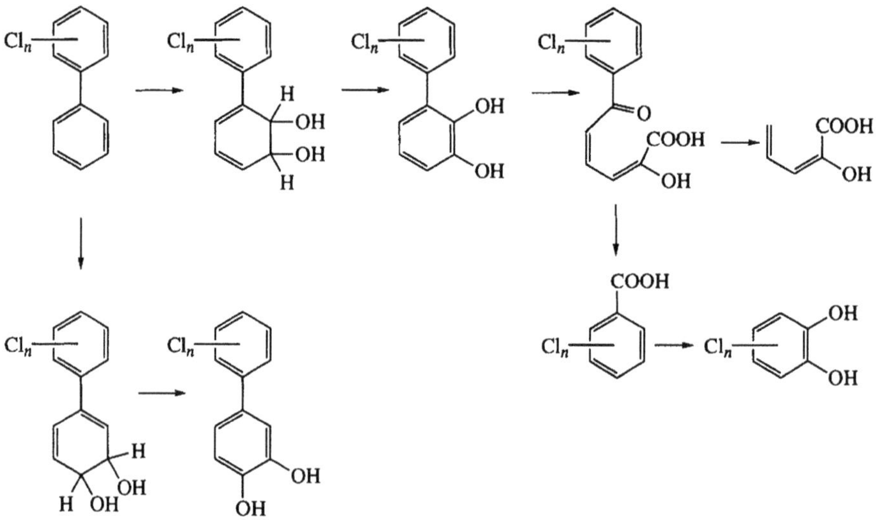

Anaerobic degradation of PCB can occur with congeners containing up to seven chlorine atoms. It generally results in *ortho*-substituted congeners, which can be further degraded aerobically.

Aerobic biodegradation of PCDDs appears to be restricted to congeners containing up to four chlorine atoms. No evidence for ring cleavage has been reported.[56] Anaerobic micro-organisms from riverine sediments reductively dechlorinate PCDDs.[57] Very few and only inconclusive reports exist about degradation of PCDFs. Some indications exist about oxygen-bridge cleavage in these compounds.[56]

Biotransformation Mechanisms in Vertebrates

Numerous foreign compounds enter a body in the course of time. This could be through inhalation, skin adsorption or, more importantly, the diet. Hydrophilic compounds can be easily excreted through bile in the faeces or through the kidney into the urine. The more lipophilic compounds are not so readily excreted and tend to accumulate in the fatty tissues of the body. This may lead to elevated and thereby toxic levels. Therefore excretion is preferable. To enhance this excretion, the molecule has to be made more hydrophilic. This process is called biotransformation.

Retention of contamination in an organism is important for two reasons: accumulation in the body leading to high body burdens, redistribution and

[57] J. E. M. Beurskens, M. Toussaint, J. de Wolf, J. M. D. van der Steen, P. C. Slot, L. C. M. Commandeur and J. R. Parsons, *Environ. Toxicol. Chem.*, 1995, **14**, 939.

possibly effects, and food-chain accumulation. Retention is governed by several factors, including oxidative metabolism, conjugation reactions and excretory mechanisms. Oxidative metabolism increases water solubility, thereby increasing blood plasm concentrations which make the compounds more readily available for urinary and biliary excretion.

Some of these oxidative metabolites are biologically active and may even be more toxic than their parent compounds. Conversion by conjugation reactions is considered to lead to termination of the toxicity of these compounds and excretion of the non-toxic conjugates.[58] Oxidative metabolism of a parent compound can lead to different products which may be excreted at different rates, as has been demonstrated for hydroxylated benzo[a]pyrenes.[58]

The metabolic potency of wildlife species with regard to CACs has been addressed initially by comparison of PCB congener patterns in different levels of food chains. In particular, the work of Boon *et al.*[20] and Tanabe *et al.*[19] on marine mammals and Walker and colleagues[59,60] on birds has been important in this respect. Typically, congener patterns were displayed by depicting the ratio of each congener analysed relative to a chosen reference congener, usually PCB-153. The reference congener is assumed to be hardly or non-degradable and indeed is usually found as the congener with the highest concentrations in vertebrates. Thus, it was possible to show that most seal species had lower ratios of the lower chlorinated congeners than the fish they preyed upon, whereas in several cetaceans these ratios were often higher than those observed in seals.[20]

Findings of actual metabolites of CACs in wildlife tissues[61,62] confirmed that biotransformation was the prime reason for certain congeners disappearing from the PCB patterns when comparing patterns in predators and prey organisms. Seven PCB congeners, *viz.* PCBs 49, 70, 87, 101, 132, 141, and 149, and also *p,p'*-DDE are consistently found to be metabolized to methylsulfonyl metabolites in fish- and seal-eating mammals.[62]

Meanwhile, in laboratory studies the metabolism of PCBs and DDE was also confirmed by detection of their metabolites.[63] Based on the many laboratory studies with terrestrial mammals (rodents mainly) and the work mentioned above, it is now believed that biotransformation occurs in three phases:

- Phase 1: activation, *i.e.* the introduction of an active group (*e.g.* an oxygen atom or a hydroxy group). As a result, the molecule becomes more water-soluble and, more importantly, more reactive.
- Phase 2: conjugation, reaction of the activated molecule with a hydrophilic molecule such as glucuronic acid, glutathione or sulfate.
- Phase 3: excretion and possibly additional reactions. The hydrophilicity of the metabolite is such that it can be transported in aqueous media like blood or bile. This enables excretion through the normal routes (kidneys and gut).

[58] J. B. Pritchard and J. R. Bend, *Environ. Health Perspect.*, 1991, **90**, 85.
[59] C. H. Walker, *Aquat. Toxicol.*, 1990, **17**, 293.
[60] J. T. Borlakoglu, J. P. G. Wilkins and C. H. Walker, *Mar. Environ. Res.*, 1988, **24**, 15.
[61] S. Jensen and B. Jansson, *Ambio*, 1976, **5**, 257.
[62] Å. Bergman, R. J. Norstrom, K. Haraguchi, H. Kiroaki and P. Béland, *Environ. Toxicol. Chem.*, 1994, **13**, 121.
[63] J. E. Bakke, Å. Bergman and G. L. Larsen, *Science*, 1982, **217**, 645.

A compound will not necessarily pass all three phases. As will be demonstrated below, a compound may well be transformed into an even more lipophilic metabolite (in contrast to the primary goal of biotransformation).

The initial activation of compounds is catalysed by various enzymes, depending on the chemical structure of the compound in question. The cytochrome P450 monooxygenase complex (CYP) is the major biotransforming system. In vertebrates it is present in all the organs of the body and particularly abundant in the liver.

The two major compounds of the complex are cytochrome P450 and NADPH cytochrome P450 reductase. The resulting overall reaction can be described as:

$$RH + O_2 + NADPH + H^+ \rightarrow ROH + H_2O + NADP^+ \tag{1}$$

CYP exhibits a great variability with respect to the type of reactions it can perform, its substrate specificity and its manifestation. Vertebrates, for example, show a striking heterogeneity in CYP forms (isozymes).[64]

A particular isozyme can perform several types of reaction, including hydroxylations, reductions, *O*-dealkylations and epoxidations.[65] The most important reactions for CAC are hydroxylations and epoxidations. A CYP isozyme can also catalyse different reactions of different substrates. CYP1A, to give an example, can *O*-dealkylate ethoxyresorufine or hydroxylate benzo[*a*]pyrene.

Substrate specificity is reflected in that each CYP isozyme requires different molecular features of its substrates. Thus, CYP2E metabolizes substrates with an alcohol or keto moiety. For CACs, CYP1A1, CYP1A2 and CYP2B are the most important isozymes. CYP1A metabolizes mostly planar substrates like 2,3,7,8-tetrachlorodibenzo-*p*-dioxin, whereas CYP2B transforms the more globular substrates, *e.g.* the transformation of the pesticide aldrin into dieldrin.

Within the group of PCB congeners, examples of both types of substrates can be found. The number and the position of the chlorines in the PCB molecule determine its three-dimensional structure and its reactivity. Four general rules for PCB metabolism have been suggested:[66,20]

1. Hydroxylation is favoured at the *para* position in the least chlorinated ring unless this site is sterically hindered.
2. Two vicinal hydrogen atoms enhance metabolism greatly.
3. An increase in chlorination at both phenyl rings leads to a decrease in the rate of metabolism.
4. The position of the hydroxy group is preferably at the *para* position or *para* to a chlorine atom.

The number of *ortho*-chlorines is important for the three-dimensional structure of the PCB. When two or more *ortho*-chlorines are present, the energy barrier that must be overcome for the molecule to become planar is relatively

[64] D. W. Nebert, D. R. Nelson, M. J. Coon, R. W. Estabrook, R. Feyereisen, Y. Fujii-Kuriyama, F. J. Gonzalez, F. P. Guengerich, I. C. Gunsalus, E. F. Johnson, J. C. Loper, R. Sato, M. R. Waterman and D. J. Waxman, *DNA Cell Biol.*, 1991, **10**, 1.

[65] J. A. Timbrell, *Principles of Biochemical Toxicology*, Taylor & Francis, London, 1982.

[66] S. Safe, in *Halogenated Biphenyls, Terphenyls, Naphthalenes, Dibenzodioxins and Related Products*, ed. R. D. Kimbrough and A. A. Jensen, Elsevier, Amsterdam, 1989, ch. 5.

large due to repulsion forces between the two approaching chlorine atoms. This has consequences for the CYP that will transform the PCB. A PCB with three or four *ortho*-chlorines has its two phenyl rings perpendicular to each other. It will most likely be transformed by the CYP form for globular substrates, *i.e.* CYP2B. A PCB that can be brought into a planar configuration easily is likely to be transformed by CYP1A.

In globular congeners the principal enzymatic reactions occur at the 3,4-positions. When planar (*i.e.* mono- or non-*ortho*-substituted) congeners are involved, this occurs usually at the 2,3-positions.[20]

Not all CYP isozymes are present in all animals. It appears that the diversity of CYPs in an animal species (but also overall CYP activity) is closely linked to its nutritional habits. As can be expected, plant-eating species like rat and chicken exhibit higher activities and a higher diversity than carnivores. Similar differences can be seen between food specialists and food generalists. Birds like eider which only eat molluscs or crustaceans show lower levels of CYP[67] than generalists, *e.g.* the herring gull.

This inhomogeneous distribution of CYP forms over the animal kingdom results in important differences in the ability to biotransform foreign compounds. This becomes obvious when studying the accumulation of PCBs in different animal phyla. On the basis of the rules given above and phenomena occurring in nature, PCBs can be divided into the four structural groups given below. This distinction is more readily applicable than the previous rules.

A. PCBs with vicinal hydrogen atoms in the *meta/para* positions and an irrelevant number of *ortho*-chlorines. These PCBs predominantly will have a globular shape.

B. PCBs with vicinal hydrogen atoms in the *ortho/meta* positions and at most one *ortho*-chlorine. These PCBs can adopt a planar configuration.

C. PCBs with vicinal hydrogen atoms in the *ortho/meta* positions and at least two *ortho*-chlorines. These PCBs will have a globular shape.

D. PCBs without vicinal H-atoms.

In rodents, group A-type substrates most likely will be transformed by CYP2B. Planar substrates will be transformed by CYP1A, whereas C- and D-type substrates will not be transformed. Hence, the structure of the compound determines which isozyme will mediate in its transformation. For other species, however, these structural rules may be different. In birds, group A-type substrates are metabolized easily, whereas planar compounds are transformed with great difficulty.

Similar species differences in biotransformation abilities can be observed in the marine environment.[20,19] Molluscs and polychaeta are not able to metabolize any PCB group. Fish seem to be able to metabolize globular but not planar congeners. In cetaceans the CYP2B isozymes seem to be relatively underdeveloped, because whale species can metabolize planar congeners only. Freshwater and other small cetaceans have an even lower capacity than larger marine species,[68,69]

[67] M. J. C. Rozemeijer, J. P. Boon, C. Swennen, A. Brouwer and A. J. Murk, *Aquat. Toxicol.*, 1995, **32**, 93.

[68] K. Kannan, S. Tanabe, R. Tatsukawa and R. K. Sinha, *Toxicol. Environ. Chem.*, 1994, **42**, 249.

[69] S. Tanabe, S. Watanabe, H. Kan and R. Tatsukawa, *Mar. Mamm. Sci.*, 1988, **4**, 103.

which may be an evolutionary result of an even less varied diet. Seals can metabolize both globular and planar CACs. The polar bear appears to possess the largest transforming capacity of marine mammals. This animal metabolizes all PCBs with two vicinal H-atoms irrespective of the number of *ortho*-chlorines. Even congeners with only a single, non-chlorinated *para* position were metabolized by the polar bear.[70] Most likely this ability is connected to the lipid-rich food and the hibernation period. During this period, the fat reserves of the animal are mobilized and metabolism is altered.

In phase 2, the activated molecule conjugates with a hydrophilic molecule. In general, this conjugated molecule is readily available *in vivo*. An important reaction is the glucuronide formation. Glucuronic acid is transferred from uridine diphosphate glucuronic acid to the active group of the compound, which is usually the epoxide or hydroxy group formed in phase 1.

A second important conjugation reaction is the sulfate conjugation. This, again, occurs predominantly with hydroxy groups and occasionally with amino groups. The sulfate donor is 3'-phosphoadenosine-5'-phosphosulfate, which is formed from ATP and inorganic sulfate. The enzyme involved is sulfotransferase.

The third major conjugation reaction is the glutathione conjugation. The enzyme glutathione-*S*-transferase catalyses the reaction of the reactive thiol group of the nucleophilic glutathione with the electrophilic centre of the phase 1 product. Subsequently, the glutamyl and glycinyl residues are removed and the cysteine moiety is acetylated (Scheme 3). Because the thiol group is so active, it can also bind to cellular proteins without the interference of the enzyme. This can lead to inactivation of the protein.[65]

A number of routes of excretion (phase 3) are available. These include routes *via* kidneys, bile, gastrointestinal tract, lung, sweat, milk, anal glands and semen. Molecular size is an important feature for excretion into the urine. Mercapturic acid pathway (MAP, see Scheme 3 and below) products can be found in urine. In general, the limit to pass the filter in the glomerulus in the kidney is a molecular weight of 300. For most animal species, conjugated metabolites of PCBs, PCDDs and PCDFs are too large to pass this filter. These metabolites are mostly excreted through the bile.

Secretion into milk is an important route of excretion. Children, for example, may take up more PCDDs and PCDFs than the acceptable daily intake through the mothers' milk.[71] Lactating female otters were the only individuals of the Dutch otter population that had PCB levels below the no-effect level for reproduction because a major part of the PCBs was excreted through the milk.

Biliary secretion, however, is the most important route for conjugated metabolites. After secretion into the gut, further metabolism and subsequent reabsorption is possible, due to an increased lipophilicity of the metabolite as is the case with methylsulfinyl and methylsulfonyl metabolites[72,63] (see Scheme 3). This may give rise to an enterohepatic circulation: (re-)absorption →modification in the liver →excretion through the bile →modification in the intestines.

[70] D. C. G. Muir, R. J. Norstrom and M. Simon, *Environ. Sci. Technol.*, 1988, **22**, 1071.

[71] H. J. Pluim, J. Wever, J. G. Koppe, J. W. van der Slikke and K. Olie, *Chemosphere*, 1993, **26**, 1947.

[72] Å. Bergman, A Biessmann, I. Brandt and J. Rafter, *Chem.-Biol. Interact.*, 1982, **40**, 123.

Scheme 3 Metabolic pathways of PCB biotransformation in vertebrates. MAP, mercapturic acid pathway; GSH, glutathione, Cys, cysteine

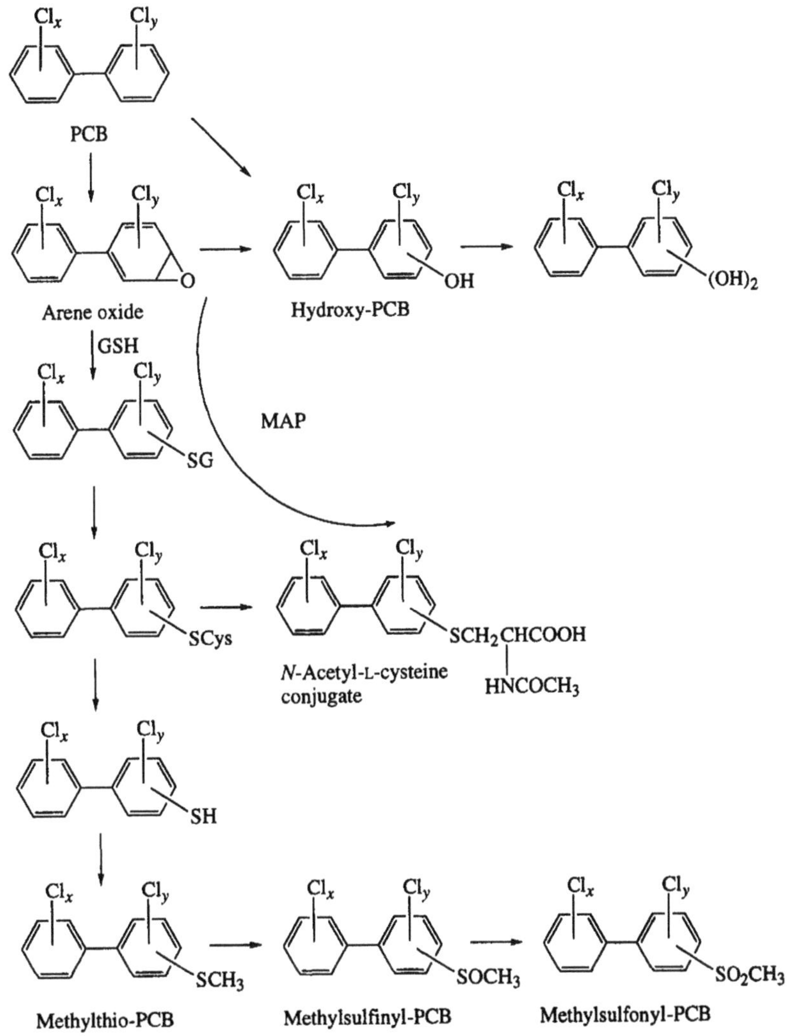

Metabolites

As mentioned above, the process of biotransformation of persistent CACs has been substantiated by the finding of metabolites in laboratory and wildlife species. In particular for PCBs, the presence of both the hydroxylated congeners in blood as well as the lipophilic methylsulfonyl-PCBs in adipose, liver and lung tissue of different species has proved that metabolism proceeds according to the mechanisms outlined above. Similarly, hydroxy metabolites of PCDFs have been found in rodents.[73]

PCBs. The metabolic pathways of PCB breakdown can thus be summarized as follows (*cf.* Scheme 3). PCBs can be transformed into arene oxides, which may be hydrolised into hydroxy- and dihydroxy-PCBs. Direct insertion of oxygen is

[73] H. Kuroki, R. Hattori, K. Haraguchi and Y. Masuda, *Chemosphere*, 1987, **16**, 1641.

another proposed mechanism[74] leading to hydroxy-PCBs. The arene oxide may also react with endogenous molecules like glutathione (GSH) to form conjugates which are excreted into the bile. These may be transformed into more hydrophilic compounds in the mercapturic acid pathway (MAP).[63,75] Cysteine (Cys) conjugates may also be metabolized by β-lyases into thiols and methyl thioethers, which can be reabsorbed from the gut and further oxidized in the liver to methylsulfinyl and methylsulfonyl compounds.[72] The latter are relatively lipophilic and are retained in adipose tissue and liver[62] and selectively in other tissues, *e.g.* lung.

PCDD/Fs. Absorption from the gastrointestinal tract of mammals is highly effective (up to more than 75%) when lower chlorinated congeners are involved, but reduces with increasing molecular size. Non-2,3,7,8-substituted PCDDs and PCDFs are readily transformed into mono- or dihydroxylated metabolites. Generalized biotransformation pathways for PCDDs and PCDFs have been presented in a recent review[30] and resemble the scheme given for PCBs. 2,3,7,8-Substituted PCDDs and PCDFs are extremely stable towards metabolic breakdown and the metabolic conversion rate of these congeners is strongly reduced. PCDFs with the 4- and 6-positions chlorinated are also persistent in biota, since these positions are favourable for enzymatic attack from an energetic point of view.

Chlorinated Pesticides. The major metabolic degradation pathways for toxaphene in all organisms are probably reductive dechlorination and reductive dehydro-chlorination. In some cases, oxidative dechlorination has been observed to result in hydroxy derivatives, acids and ketones.[76] Aldrin is transformed into dieldrin in biotic as well as abiotic transformation processes.

The major breakdown products of DDT are DDE and DDD. These products can be further degraded to more hydrophilic compounds like dichlorodiphenylacetic acid (DDA). Stable lipophilic metabolites of *p,p'*-DDE have also been found, such as the phenolic metabolites[77] and the *ortho-* and *meta*-methylsulfonyl-DDEs. In particular, the latter have been detected in many wildlife species since the 1970s.[61] Major breakdown pathways of DDT are depicted in Scheme 4.

4 Effects

Several recent reviews have concluded that, until now, no definitive cause–effect relationships have been established between exposure to environmental concentrations of chlorinated chemicals and serious adverse health consequences for humans.[78–80]. For fish-eating birds and mammals, however, substantial evidence

[74] L. T. Burka, T. M. Plucinski and T. L. Macdonalds, *Proc. Natl. Acad. Sci. USA*, 1983, **80**, 6680.

[75] N. P. E. Vermeulen, J. de Jong, E. J. C. van Bergen and R. T. H. van Welie, *Arch. Toxicol.*, 1989, **63**, 173.

[76] M. A. Saleh, *Rev. Environ. Contam. Toxicol.*, 1991, **118**, 1.

[77] G. Sundström, B. Jansson and S. Jensen, *Nature*, 1975, **255**, 627.

[78] M. A. Kamrin and L. J. Fischer, *Environ. Health Perspect.*, 1991, **91**, 157.

[79] R. F. Willes, E. R. Nestman, P. A. Miller, J. C. Orr and I. C. Munro, *Regul. Toxicol. Pharmacol.*, 1993, **18**, 313.

[80] E. Delzell, M. D. Doull, J. P. Giesy, D. Mackay, I. C. Munro and G. Williams, *Regul. Toxicol. Pharmacol.*, 1994, **20**, 1.

Scheme 4 Major pathways of *p,p'*-DDT biotransformation

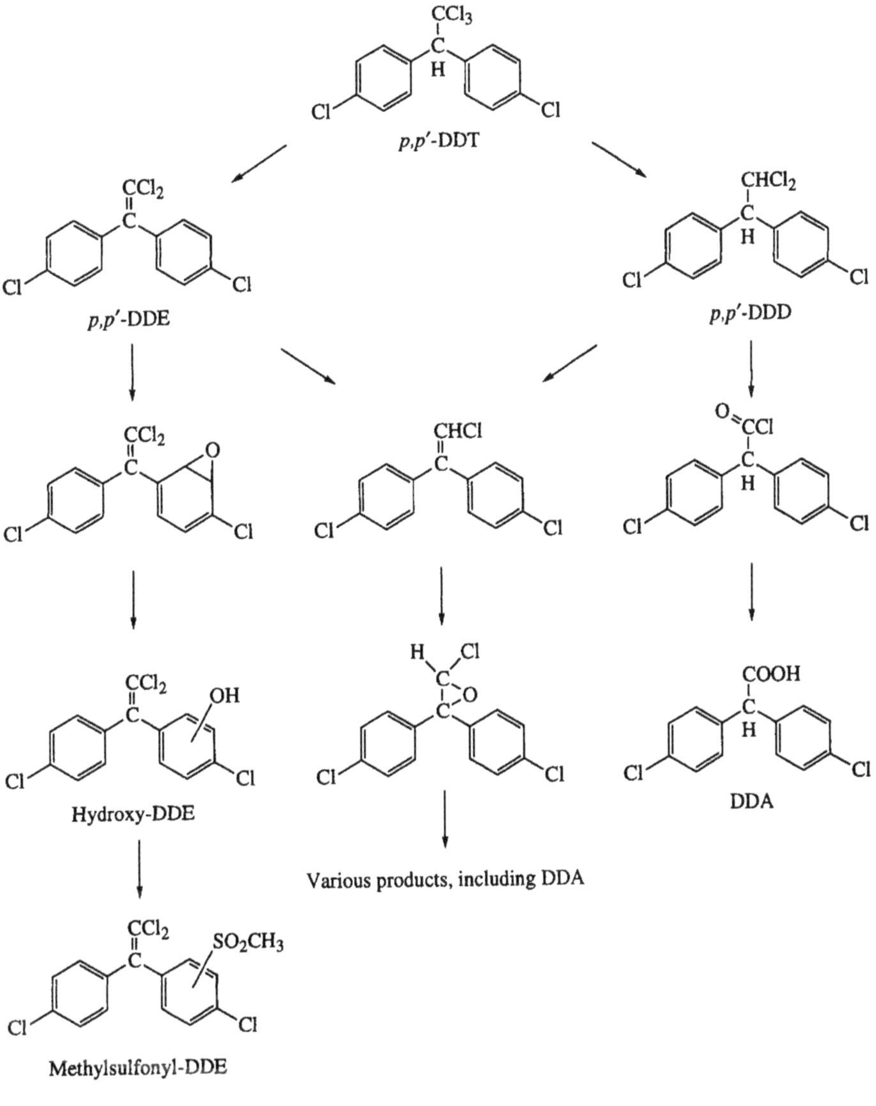

has now been provided in the literature that they are threatened when living in relatively contaminated areas. The most critical toxicants for causing birth effects and embryo lethality are the planar CACs.[81,82]

Experimental Studies

In laboratory vertebrates, which are mostly terrestrial, the biological effects of CACs observed include hepatic damage, dermal disorders, reproductive toxicity, thymic atrophy, body weight loss, immunotoxicity, endocrine disruptions and

[81] M. Gilbertson, T. Kubiak, J. Ludwig and G. Fox, *J. Toxicol. Environ. Health*, 1991, **33**, 455.
[82] J. P. Ludwig, J. P. Giesy, C. L. Summer, W. Bowerman, R. Aulerich, S. Bursian, H. J. Auman, P. D. Jones, L. L. Wiliams, D. E. Tillitt and M. Gilbertson, *J. Great Lakes Res.*, 1993, **19**, 789.

teratogenicity. While the toxicity of PCDDs and PCDFs is produced by the parent compounds[30] and only observed for 2,3,7,8-substituted compounds, for PCBs the situation is more complicated and highly dependent on the structural molecular features. Parent PCBs may exert effects similar to PCDDs and PCDF when their chlorine substitution pattern permits a quasi-planar structure (similar to that of the various planar PCDD and PCDF molecules) to be attained relatively easily. In addition, for this 'dioxin-type' toxicity, chlorine substitution in two *para* positions and at least two *meta* positions relative to the pivot bond is necessary. All non-*ortho*-, several mono-*ortho*- and a few di-*ortho*-substituted PCBs fulfilling these criteria exhibit dioxin-type effects. These effects are mediated through the cellular Ah-receptor. Interaction of planar CACs with this receptor results in the formation of a receptor–substrate complex, which is translocated to the cell nucleus. Here, binding to dioxin responsive elements on the DNA occurs, which leads to subsequent induction of CYP1A isozymes as well as several toxic responses.[83] The latter include immunotoxicity, teratogenicity, carcinogenicity and porphyria.[84]

Because all 2,3,7,8-substituted PCDDs and PCDFs, as well as the planar PCBs, elicit this type of Ah-receptor mediated responses, their toxicity can be expressed relative to that of the most potent congener, which is 2,3,7,8-tetrachlorodibenzo-*p*-dioxin (TCDD). This concept is known as the toxic equivalence factor (TEF) approach.[84,13] The TEF concept can be used to classify each individual congener and, by an additive approach, assess the total risk of environmental or biological levels of PCDDs, PCDFs and planar PCBs. The predictive value of this approach, in particular when PCBs are included, appears to be species- as well as response-dependent.[13] This is primarily due to the presence of other PCBs in environmental residues, which may act antagonistically.

Planar PCBs, in particular PCBs 105, 118, 126 and 156,[8] contribute quite significantly to the total toxic equivalent (TEQ) values detected in wildlife. This contribution usually exceeds 50% and can be as high as 90% in aquatic species. The remaining contribution is roughly 3–4:1 PCDD:PCDFs.[85,86] In terrestrial species and omnivores, the PCB contribution is significantly lower than in aquatic species. This is most probably a result of the higher potential of terrestrial animals to degrade PCBs, the larger extent of bioaccumulation of PCBs in aquatic foodchains and the different sources of contamination in both environments.

PCBs with a more globular structure elicit effects similar to phenobarbital. These include induction of the isozyme CYP2B, carcinogenic promoter activity[87] and neurotoxicity. Developmental neurotoxicity of PCBs in animals is consistent with findings in children and results in persistent behavioural and neurological effects, notably alterations in motor development and cognitive function.[88] The

[83] J. P. Landers and N. J. Bunce, *Biochem. J.*, 1991, **276**, 273.

[84] S. H. Safe, *Crit. Rev. Toxicol.*, 1990, **21**, 51.

[85] J. P. Giesy, J. P. Ludwig and D. E. Tillitt, *Environ. Sci. Technol.*, 1994, **28**, 134A.

[86] J. T. Sanderson, R. J. Norstrom, J. E. Elliott, L. E. Hart, K. M. Cheng and G. D. Bellward, *J. Toxicol. Environ. Health*, 1994, **41**, 247.

[87] E. M. Silberhorn, H. P. Glauert and L. W. Robertson, *Crit. Rev. Toxicol.*, 1990, **20**, 439.

[88] H. A. Tilson and G. J. Harry, in *The Vulnerable Brain and Environmental Risks*, eds. R. L. Isaacson and K. F. Jensen, Plenum Press, New York, 1994, ch. 14.

precise mechanism of this neurotoxicity remains to be elucidated.

Metabolites of PCBs also exert biological effects. The Ah-receptor mediated responses, however, are probably caused by the parent compounds only. The effects of hydroxylated PCBs include inhibition of cytochrome P450-dependent enzyme activities and competitive interference with thyroid hormone and vitamin A metabolism.[89,90] Methylsulfonyl-PCBs have been shown to inhibit aryl hydrocarbon hydroxylase activity[91] and to elicit phenobarbital-type toxicity and may in fact be responsible for the observed effect presumed to be caused by the parent compound.[92]

Field Observations

Effects in feral populations are not only much more difficult to observe and ascertain, but also very unlikely to be caused by a single cause, like pollution by a single group of contaminants. Many other factors can play a role, such as temperature, population densities and toxic algal blooms, which impede efforts to link causally the effects observed to pollution. They may also be caused by a combined effect of the presence of PCBs, PCDD/Fs, DDTs, drins and HCB.[80]

Yet, quite a few effects that have been observed in wildlife are highly similar to those observed in controlled laboratory exposure studies. Recent experiments with wildlife species under controlled exposure conditions have shown that some of the effects observed in the field can indeed at least partly be ascribed to environmental contamination with CACs.

Seals and Other Marine Mammals. Among the effects of CACs reported on free-ranging seals are skeletal deformities and impaired reproduction. Captive seals fed with CAC contaminated fish showed a significantly reduced reproductive success compared to animals fed with clean fish.[93] Vitamin A and thyroid hormone levels were also significantly reduced.[89] These findings corresponded to results of experimental studies with mink and rodents exposed to PCBs.

Recent experimental studies with seals conducted in The Netherlands have shown that immunosuppression in these animals is also causally linked to exposure to PCBs.[94,95] These studies demonstrated that chronic exposure by CACs taken up from contaminated fish affected immune function significantly, whereas relatively short-term fasting periods do not seem to pose additional risks. It was concluded that environmental contaminants will adversely affect immune function in free-ranging seals.[96] The importance of these findings must

[89] A. Brouwer, P. J. H. Reijnders and J. H. Koeman, *Aquat. Toxicol.*, 1989, **15**, 99.

[90] A. Brouwer, E. Klasson-Wehler, M. Bokdam, D. C. Morse and W. A. Traag, *Chemosphere*, 1990, **20**, 1257.

[91] J. Lund, Å. Bergman and I. Brandt, *Toxicol. Lett.*, 1986, **32**, 261.

[92] Y. Kato, K. Haraguchi, M. Kawashima, S. Yamada, Y. Masuda and R. Kimura, *Chem.-Biol. Interact.*, 1995, **95**, 257.

[93] P. J. H. Reijnders, *Nature*, 1987, **324**, 456.

[94] P. S. Ross, R. L. de Swart, P. J. H. Reijnders, H. van Loveren, J. G. Vos and A. D. M. E. Osterhaus, *Environ. Health Perspect.*, 1995, **103**, 162.

[95] R. L. de Swart, P. S. Ross, H. H. Timmerman, H. W. Vos, P. J. H. Reijnders, J. G. Vos and A. D. M. E. Osterhaus, *Clin. Exp. Immunol.*, 1995, **101**, 480.

[96] R. de Swart, Ph.D. Thesis, University of Rotterdam, 1995.

be underlined, given the mass mortalities that have occurred in populations of pinnipeds and cetaceans in European and Pacific coastal waters.

A number of other biological effects have been observed in pinnipeds and cetaceans and have been summarized.[97] Epidemiological evidence of contaminant-induced effects included abortions, tumours, decreased fecundity, impaired reproduction and reduced testosterone levels.

Birds. The last decades have shown a vast and still increasing number of both experimental and field studies to assess the importance of current environmental exposure levels of CACs for reproduction, teratogenic (developmental deformities) and other effects in birds.[98,99] The major efforts in this field have been the ongoing field studies in the Great Lakes piscivorous birds and their prey. These studies have been reviewed in several papers[81,85] and conclude that piscivorous birds from this area display symptoms of exposure to chlorinated chemicals, which, although their concentrations decline, still cause embryo lethality or deformities in all populations investigated. Symptoms resulting from controlled laboratory studies with multiple compound exposure[100-102] have also been observed in field populations.[81,103-105] These include biochemical responses: enzyme induction, vitamin A and thyroid hormone regulation, as well as decreased breeding success, egg shell thinning, chick edema disease, growth retardation, bill deformities and neurotoxic responses.

Similar findings have been reported on a more incidental base from other areas in the world. From these, the common estuaries of the rivers Rhine/Meuse/Scheldt in Northwest Europe are relatively well studied with regard to chlorinated aromatic compounds. Breeding success of several birds species in this area (cormorant and common tern) has been shown to be related to contamination levels.[106,107] Other species, however, do not show significantly reduced breeding success and do not seem to be at risk.[108]

[97] P. S. Ross, I. K. G. Visser, H. W. J. Broeders, M. van de Bildt, D. Bowen and A. D. M. E. Osterhaus, *Vet. Record*, 1992, **130**, 514.

[98] N. Yamashita, S. Tanabe, J. P. Ludwig, H. Kurita, M. E. Ludwig and R. Tatsukawa, *Environ. Pollut.*, 1993, **79**, 163.

[99] J. D. Hutchinson and M. P. Simmonds, *Rev. Environ. Contam. Toxicol.*, 1994, **136**, 82.

[100] L. E. Hart, K. M. Cheng, P. E. Whitehead, R. M. Shaw, R. J. Lewis, S. R. Ruchowski, R. W. Blair, D. C. Bennett, S. M. Bandiera, R. J. Norstrom and G. D. Bellward, *J. Toxicol. Environ. Health*, 1991, **32**, 331.

[101] A. J. Murk, A. T. C. Bosveld, A. Barua, M. van den Berg and A. Brouwer, *Aquat. Toxicol.*, 1994, **30**, 91.

[102] M. van den Berg, B. L. H. J. Craane, T. Sinnige, S. van Mourik, S. Dirksen, T. J. Boudewijn, M. van der Gaag, I. J. Lutke-Schipholt, B. Spenkelink and A. Brouwer, *Environ. Toxicol. Chem.*, 1994, **13**, 803.

[103] T. J. Kubiak, H. J. Harris, L. M. Smith, T. R. Schwartz, D. J. Stalling, J. A. Trick, L. Sileo, D. E. Docherty and T. C. Erdman, *Arch. Environ. Contam. Toxicol.*, 1989, **18**, 706.

[104] T. Colborn, *J. Toxicol. Environ. Health*, 1991, **33**, 395.

[105] J. E. Elliott, R. W. Butler, R. J. Norstrom and P. E. Whitehead, *Environ. Pollut.*, 1989, **59**, 91.

[106] S. Dirksen, T. J. Boudewijn, L. K. Slager, R. G. Mes, M. J. M. van Schaick and P. de Voogt, *Environ. Pollut.*, 1995, **88**, 119.

[107] A. T. C. Bosveld, J. Gradener, A. J. Murk, A. Brouwer, M. van Kampen, E. H. G. Evers and M. van den Berg, *Environ. Toxicol. Chem.*, 1995, **14**, 99.

[108] J. Stronkhorst, T. J. Ysebaert, F. Smedes, P. L. Meininger, S. Dirksen and T. J. Boudewijn, *Mar. Pollut. Bull.*, 1993, **26**, 572.

Eggshell thinning has been known, since the 1960s, to be caused by pesticides, in particular *p,p'*-DDE.[109] Many later reports have substantiated this relationship, although it is likely that other compounds present in the field, such as the PCBs, may also relate to this effect.[106]

A comparison of concentrations in eggs of free-ranging birds with the lowest observed effect levels for reproduction or deformities has been presented recently.[46] It was concluded that adverse effects may be expected from PCBs for herring gull, great blue heron, Forster's tern and common tern at several locations in the world.

Biomarkers

Although the subject of biomarkers is outside the scope of this study, the above discussion has shown that biochemical effects are resulting from exposure to CACs. Some of these have been advocated as possible screening methods for exposure to PCBs, PCDDs and PCDFs.[110] Among these are several *in vivo* and *in vitro* assays to assess induction of CYP isozymes like CYP1A1, as well as vitamin A and thyroid hormone levels. CYP1A1 induction, as we have seen specific for planar compounds,[111] can be determined by measuring the activities of 7-ethoxyresorufin *O*-deethylase (EROD) or aryl hydrocarbon hydroxylase (AHH). These bioassays are very sensitive, cost and time effective and can serve to limit the number of expensive congener-specific analyses.

Perhaps the most important argument to use such biomarker approaches lies in their inherent potential to integrate the exposure to compounds that are hitherto unknown or unidentified and to account for interactions like synergism or antagonism.

5 Conclusions

The knowledge of ecotoxicology of chlorinated aromatic compounds has increased substantially over the last 10 years. In particular, observations in wildlife have shown that, contrary to what is believed now for humans, certain PCBs, PCDDs and PCDFs are causing harmful effects to individuals and populations, despite the fact that their concentrations in both the abiotic and the biotic environment have been declining continuously since the 1970s.

Although fish-eating birds possess the ability to metabolize CACs, the relative amounts of metabolites found in their tissues is lower than that observed in marine mammals. Apparently their metabolic capacity is less than that of pinnipeds and probably also less than that of some cetaceans. Cetaceans, in turn, have a lower metabolic capacity than pinnipeds and may therefore be more at risk.

Terrestrial mammals have lower levels of PCBs, PCDDs and PCDFs than aquatic species or piscivorous birds. This is most probably caused by a lower level of exposure (through the food) and a higher metabolic capacity.

Chlorinated aromatic compounds are dispersed globally because of their

[109] D. A. Ratcliffe, *Nature*, 1967, **215**, 208.
[110] A. T. C. Bosveld and M. van den Berg, *Fresenius' J. Anal. Chem.*, 1994, **348**, 106.
[111] T. D. Bucheli and K. Fent, *Crit. Rev. Environ. Sci. Technol.*, 1995, **25**, 201.

persistence and their long-range transport, and the acting of oceans as their sinks. As a result of the latter phenomenon, the aquatic environment is the most important compartment from an ecotoxicological point of view. Adverse effects on wildlife species are more likely to occur in aquatic oriented than in terrestrial species.

Despite their persistency, chlorinated aromatic compounds can be transformed by microflora and fauna as well as higher organisms. The products of these processes are different. Metabolism is an important process for vertebrates to regulate or detoxify their body burden of these compounds. Some products of the biotransformation process, however, may elicit toxic effects or accumulate themselves, thereby acting opposite to the goals of biotransformation.

6 Acknowledgement

The author expresses his gratitude to Marcel Rozemeijer and Heleen Goorissen for the stimulating discussions which helped to build and shape this article.

Chlorinated Pesticides

ALAN TURNBULL

1 Introduction

Along with some other chlorinated micropollutants, persistent pesticides such as DDT have become widely distributed around the planet. Pesticides are unique in that they are purposefully released into the environment to kill selected species. The chlorinated pesticides are now largely banned chemicals which may still pose a threat to human health as well as the wider environment. Hailed as a miracle of science upon their introduction during the 1930s to 1950s, they are now under suspicion as possible factors in various human cancers and infertility problems.[1] This chapter will briefly outline the history of their introduction and subsequent restriction, along with a review of environmental distribution and transport mechanisms. A selection of data from the literature will reflect the levels detected in differing environmental media. Finally, an insight into some of the potential health and ecological problems that have been identified will be outlined.

2 A Brief History of the Organochlorine Pesticides

At the beginning of the twentieth century, early investigations with chemical pesticides led to the widespread use of inorganic compounds within agriculture containing elements such as sulfur, arsenic, mercury, lead and other metals. Some natural products, such as pyrethrum, were also known to be effective pesticides at this time, but were considered too expensive for widespread use. Between the world wars, the development of the chlor-alkali industry provided the raw material for the mass production of synthetic chlorinated organic molecules. With the increasing desire for selective biocides within agriculture, and the emerging chemical technologies, the hunt for synthetic pesticides was on.

An early chlorinated phenoxy acid herbicide (2,4-D) was first discovered in 1932. Although this compound rapidly breaks down in the environment, the seed fungicide hexachlorobenzene (HCB), introduced in 1933, was found to be far more persistent.[2] The structurally similar insecticide hexachlorocyclohexane or

---

[1] E. Carlsen, A. Giwercman, N. Keiding and N. E. Skakkerbaek, *Environ. Health Perspect.* 1995, **103**, sup. 7, 137.

[2] M. A. S. Burton and B. G. Bennett, *Sci. Total Environ.* 1987, **66**, 137.

HCH (historically known as benzene hexachloride – BHC) also appeared at this time. The outbreak of war in 1939, and the need to control malaria and typhus amongst troops and civilians, led to the discovery, development and application of DDT across the world within $4\frac{1}{2}$ years from 1940. Parallel research into nerve gas agents in Nazi Germany led to the discovery of the related organophosphorus pesticides. By the end of the war, a bright new future for agrochemical control using these organochlorine compounds was envisaged.

Following the war the UK government envisaged a practical need to modernize agriculture and increase food production by the introduction of more complex machinery, creation of larger fields, use of chemical fertilizers and the new synthetic pesticides. By 1953, two insecticidal seed dressings, dieldrin and aldrin, were being introduced into the UK. In America, toxaphene was first produced in 1945 as an effective insecticide for cotton plants. This mixture of over 170 chlorinated boranes known as camphachlor in Europe, was recommended as an alternative to DDT before it too was banned in the 1980s due to its environmental toxicity.

These new biocides were initially hailed as a miracle of modern scientific technology which would help heal the increasing post-war food problems. History now partially charts the nature and extent of the effects with which these compounds have become associated. Rachel Carson, considered to be one of the first environmentalists, commanded public attention in 1962 with her seminal text *Silent Spring*, in which she highlighted the rapid bioaccumulation and acute toxic effects these organochlorine pesticides were having on American and European rural wildlife.[3] Species such as the peregrine falcon approached extinction in much of England prior to government action, and there were many thousands of bird and mammal deaths reported across the country. The unfolding of events in the UK, and the role of the Nature Conservancy in advising increased regulation, have been reviewed by Sheail.[4] Perhaps of equal significance to the agricultural community was the emerging reality of biological resistance within target organisms. It is now understood that considerable care must be employed to avoid over zealous use of individual compounds and resulting pest resistance.

Along with other persistent organochlorines such as chlorinated biphenyls, chlorofluorocarbons, dibenzodioxins and dibenzofurans, the chlorinated pesticides have the potential to cause significant damage to the natural ecosystem by interfering with reproductive processes, thus influencing the biodiversity of non-target organisms, as well as allowing resistant gene pools within pest communities to dominate. Some aspects of this damage are now well documented, some aspects are in remission and others are perhaps as yet reserved only for pseudo-scientific speculation.

In relation to this article, those chlorinated pesticides which are rapidly degraded in the environment, and considered as non-persistent, will not be discussed. Compounds such as atrazine, chlorpyrifos and the phenoxy acid herbicides (2,4-D), mecoprop, MCPA, *etc.*) all contain organic chlorine and may

[3] R. Carson, *Silent Spring*, Penguin, London, 1962.

[4] J. Sheail, *Pesticides and Nature Conservation, the British Experience 1950–75*, Clarendon Press, Oxford, 1985.

114

be associated with the relatively short-term contamination of water, soil or food. Clearly this does not mean they pose no risk, but that the risk is likely to be localized, and environmental effects will subside once the active ingredient has been degraded to its less harmful chemical constituents. However, the persistent chlorinated insecticides may now be considered as ubiquitous global pollutants whose environmental effects will evidently only slowly diminish. Their common character can be seen in the global distribution which is now known to be due largely to long-range atmospheric distribution, predominantly in the vapour phase. These airborne residues attempt to reach an equilibrium with other environmental media such as surface waters, suspended particulate material and biological tissues. Whether from historic application in the West or from continuing current use in the developing world, these compounds can now be detected in the most remote regions of the planet.

3 Organochlorine Pesticide Production, Use and Regulation

Following the early scientific and public protests during the 1960s, many organochlorines were identified as presenting an unacceptable risk to human health and the wider environment. Scientific research has done much to understand pesticide mobility, bioaccumulation and toxicity within ecosystems throughout the planet. This has encouraged political will in the West to prohibit the retail or use of many of the more dangerous compounds. In Europe, a Council Directive of December 1978 established a framework for their control along with certain other plant protection products.[5] The UK also highlights many of these compounds on its 'Red List' of priority pollutants in surface waters.[6] Along with some other countries, the UK government monitors and enforces control limits on organochlorine residues in drinking water and in imported and regional food supplies.[7] The compounds covered and the dates of final restrictions on the sale and use in Europe are given in Table 1. Similar restrictions are now beginning to be achieved in developing countries, but economic and political problems mean that many years will probably be required to eradicate their use in these regions.

With the restriction of most organochlorine pesticides within the developed world between 1970 and 1980, the major release of the compounds in Table 1 now relates to use in developing countries, especially Asia, South/Central America, China and Africa, although statistics on their use in many areas remain unclear.[8] The following briefly reviews the major types of persistent organochlorine pesticides found in the wider environment. The structures of a selection of these are illustrated in Figure 1.

Hexachlorohexanes

These compounds were initially produced as an isomeric mixture of chlorinated

[5] EEC directive prohibiting the placing on the market and use of plant protection products containing certain active substances (79/117/EEC), 1978.
[6] A. R. Agg and T. F. Zabel, *J. IWEM*, 1994, **4**, 44.
[7] MAFF, *Report of the Working Party on Pesticide Residues: 1988–90*, HMSO, London, 1992.
[8] FAO, *FAO Yearbook – Production*, Food & Agriculture Organisation of the United Nations, Rome, 1989, vol. 43.

Table 1 Dates of final restriction of some of the organochlorine pesticides

Compound	Date restricted by European directive[a]
Aldrin	1/1/91
Endrin	1/1/91
Dieldrin	1/1/81
DDT	1/1/86
HCH (<99%γ-HCH)	1/1/81
HCB	1/1/81
Chlordane	1/1/81
Toxaphene	1984
Heptachlor	1/1/81

[a]Directive prohibiting the placing on the market and use of plant protection products containing certain active substances (79/117/EEC), 1978.

Figure 1 Structure of some of the organochlorine pesticides

Lindane   Endrin   Toxaphene congener

DDE   HCB   DDT

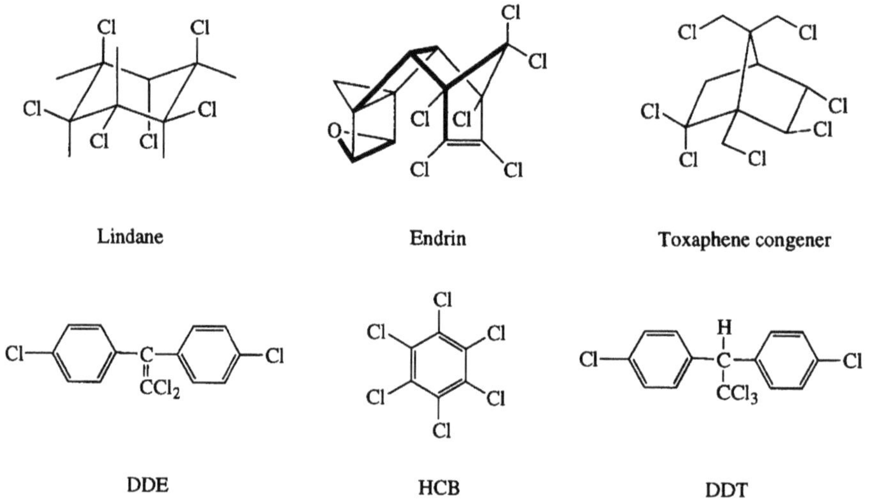

cyclohexanes known as technical HCH with the following percentages; α, 55–70%; β, 5–14%; γ, 10–18%; δ, 6–10%; and ε, 3–4%. It was subsequently discovered that only the γ isomer, known as lindane, had any insecticidal activity, which led to technical HCH being restricted across the western world along with the other organochlorines. However, lindane remains licensed for use in all but a few countries, with UK agricultural release reported to be in excess of 60 tonnes annually.[9] HCH (or lindane) is used for insect control in agriculture and forestry, although it also has extensive uses for wood preservation, public health and human medicine. Technical HCH remains a common compound used in large quantities in India, China, Africa and South America. Cumulative use data for HCH are scarce and often unreliable, but India has recorded an annual use of around 25 000 tonnes over recent years, and one factory in China is thought to have an annual production of 20 000 tonnes. Japan is reported to have consumed 400 000 tonnes between 1948 and 1970.[10] Other noted current consumers include

[9] R. P. Davis, M. R. Thomas, D. G. Garthwaite and H. M. Bowen, *Pesticide Usage Survey Report* 108: *Arable Farm Crops in Great Britain*, MAFF, London, 1992.
[10] J. M. Pacyna and M. Oehme, *Atmos. Environ.*, 1988, **22**, 243.

Italy and Argentina, although most other countries still use lindane or HCH to some extent.

Hexachlorobenzene

Current use of HCB has declined rapidly over recent years from its past use as a fungicide on cereal grains. Annual global production has been estimated at 10 000 tonnes per year in 1986,[2] although India alone is reported by the FAO to have applied 37 000 tonnes in 1988. Other sources of HCB are known to include industrial emissions, especially from the manufacture of poly(vinyl chloride), and it has also been reported as a biotransformation product of lindane in the environment. Its high persistence and volatility have combined to make HCB one of the most evenly distributed global pollutants, with levels of around 100 pg m^{-3}.

DDT

Dichlorodiphenyltrichloroethane (DDT) is perhaps the most well recognized of the persistent organochlorine insecticides and remains in use for the control of typhoid and malaria in tropical regions. Subsequent to release, various associated products are created biologically or chemically. Thus DDT levels are often given as the sum of various isomers of DDT, DDE and DDD, with DDE being generally the most stable and predominant compound, especially in biological tissue. Irrespective of the known environmental problems and extensive restrictions imposed in the West, large quantities are still used in the warmer climates of the developing world. Some foresee a continuing need for DDT use in the tropic for some years to come, and it is likely the application in such areas even now exceeds the initial use in the West. This has led to the global occurrence of DDT and its metabolites shifting from northern to more southerly latitudes.[11]

Chlorocyclodienes (Dieldrin, Aldrin, Endrin)

These pesticides act generally as persistent contact stomach acting insecticides. Dieldrin and aldrin were developed following the war and were extensively used in Europe during the 1950s and 1960s. They were used in sheep dips and as a seed dressing in the UK, although they have continuing uses for the control of tropical diseases derived from mosquitoes and locusts. It was these compounds, applied as seed dressings in the UK, which were implicated in the large numbers of bird deaths which were significant in motivating regulatory action in the 1960s. Endrin has been used as a foliar insecticide on a wide variety of crops and is given as an example of this group of compounds in Figure 1. Since their decline in use and subsequent ban in most developed countries, current application appears to be restricted to selected areas of Africa, Asia, South and Central America. Global production data for these compounds are scarce although their occurrence in remote regions is indicative of considerable use, and a behaviour similar to that of

[11] H. Iwata, S. Tanabe, N. R. Sakal and R. Tatsukawa, *Environ. Sci. Technol.*, 1993, **27**, 1080.

the other organochlorines. There are generally less environmental data for these compounds than for HCB, DDT and HCHs.

Toxaphene

Toxaphene is the collective name for around 200 congeners from the chlorination of camphene. A single constituent with seven chlorine substitutions is illustrated in Figure 1, although number of substitutions ranges from around 4 to 10 chlorines. It has not been used extensively in Europe, the main application having been the control of insect pests on cotton in the USA, and to control unwanted fish species in the North American lakes. Other major users have included Egypt, Brazil and the former Soviet Union, who have all now restricted its use. Some areas, such as Argentina and Mexico, retain legitimate use of toxaphene, although most countries now have some form of restriction on its application. Early global production estimates were put at around 400 000 tonnes, but further investigation has revealed that a true value is likely to be in excess of 1.3 million tonnes.[12] Owing to its complex mix of congeners, and its environmental behaviour, the problems and occurrence of toxaphene have been likened to those of the chlorinated biphenyls (PCBs).

Chlordane

Chlordane, like toxaphene, comprises a number of compounds of which the most generally reported in the environment are *cis/trans*-chlordane and *cis/trans*-nonachlor. Its uses include the control of termites, cockroaches and wasps. Environmental concentrations are less commonly reported for chlordane and production data are not easily available.

Analytical Considerations

It was perhaps fortunate that analytical techniques suitable for the detection of very low levels of these organochlorines were developed earlier than for many other compounds. Their stability, semi-volatile character and high degree of chlorination made the use of gas chromatography with the halogen-selective electron capture detector (GC–ECD) relatively simple and highly efficient. Because of this, and the significant toxicological effects, a large quantity of analytical data on the organochlorines has been published since their initial release.

4 Partitioning in the Environment

The persistent organochlorine compounds, once released, will partition between environmental media according to their physical and chemical properties. Steady state equilibrium partitioning between these media has been considered as the simplest model simulating their behaviour. Over recent years, modelling the environmental partitioning and fate of these compounds has led to a broad

---

[12] E. C. Voldner and Y. F. Li, *Chemosphere*, 1993, **27**, 2073.

understanding of how organochlorines have become mobilized and globally distributed. Clearly, in the real environment, partitioning is a highly complex combination of forces, only ever reaching partial equilibrium. Experimentally determined and theoretically defined partition coefficients between different media form the basis by which actual environmental behaviour is often described. Careful measurement of actual environmental levels provides the essential evidence by which all complex and simplistic modelling exercises are validated.

Partitioning of each compound will occur between adjacent media in an attempt to reach a stable equilibrium according to its relative 'solubility' in each medium. Acting alongside this process will be external and dynamic forces which influence those media. For example, a field-applied pesticide might slowly dissolve into soil water which may then be advected by hydrological forces into surface or ground water following heavy rainfall. The mobility of this compound would therefore depend upon the absolute solubility, the rate at which it is dissolved, and the dynamics of the hydrological regime driving the water movement. For most modern pesticides, the relative instability suggests transport far from the site of application is unlikely due to insufficient resistance to degradation mechanisms such as photolysis, hydrolysis or microbial decomposition. For the hydrophobic semi-volatile organochlorines, water solubility is relatively low and decomposition rates may be very slow. This results in preferential accumulation onto biological particles, particularly organic detritus, and subsequent incorporation into fatty tissues throughout the various natural food chains. In addition, volatilization followed by atmospheric transport provides the rapid and comprehensive mechanism for their distribution around the planet. These two processes have combined to allow organochlorines to accumulate both in local food chains and in areas far from the sites of original application. An overall picture of distribution would reveal a continental source, followed by atmospheric transport to a marine or polar region, and final sedimentation into the ocean floors or polar snowpacks. Over recent years it has become clear that long-range atmospheric transport leads to this preferential accumulation in cooler polar regions, which has been given the terms 'cold condensation' or 'global fractionation'.[13,14] The patterns of this global distribution are influenced by factors such as seasonal shifts in meteorology, original sites and amounts of release, and the inherent properties of the individual compounds. Whilst on their 'journey', biological tissue provides a long-term storage medium for these pesticides.

The relative persistence of individual compounds will clearly determine their longer term impact around the planet. Such data are usually provided by their experimental half-lives. However, there are very large variations in this value for a given molecule in different media and under differing soil, water or meteorological conditions. Table 2 gives a general view of the relative persistence of each compound in four major environmental compartments, although these values will reflect loss mechanisms such as leaching and evaporation in addition to decomposition. Overall, persistence in the environment increases moving from air to water to soils to sediments. However, such increases in persistence may be small, as in the case of the very stable HCB and the DDT molecules, or much

[13] F. Wania and D. Mackay, *Ambio*, 1993, **22**, 10.
[14] W. E. Cotham and T. F. Bidleman, *Chemosphere*, 1991, **22**, 165.

Table 2 Approximate
half-lives of the
organochlorine pesticides
in the environment

	Endrin	*Lindane*	*HCB*	*DDT*
Air	5 hours	2 days	2 years	2 years
Water	3 weeks	3 weeks	6 years	6 years
Soil	2 months	2 months	6 years	6 years
Sediment	8 months	8 months	6 years	6 years

greater as in the case of aldrin, endrin and dieldrin which are far less stable in the atmosphere than in soils and sediments.

To understand the mechanisms of pesticide behaviour in the environment, simple relationships and theoretical models have been derived to describe partitioning in a myriad of field and laboratory experiments. Some of the more important areas of research are briefly discussed below.

Site of Application

Environmental fate now has a high priority within the agrochemical industry and researchers extensively investigate the immediate impact of agricultural pesticide applications. It is known that only a very small percentage of the active ingredient ever reaches target organisms. The remainder will either degrade, be tied up in the soil, be released to surface water or evaporate into the atmosphere. A recent UK project has shown that losses to water for all pesticides only constitute a minor fraction of mass removal, typically less than 1%.[15] In this work, it was determined that the hydrological regime was the predominant factor influencing pesticide concentrations in runoff water. However, the resultant transient aqueous concentrations following storm rainfall may not be trivial, and current UK agricultural standards on pesticide use have been shown to lead to acute toxic effects in surface waters adjacent to sites of application under extreme, although not unusual, conditions.[16] General reviews of pesticide mobility from agricultural land and the potential effects on water quality have appeared in the literature.[17] For those hydrophobic compounds which predominantly adsorb to particles, water borne losses may be considerably less than 0.1% of that applied. The degree of adsorption to soil or suspended sediments, the partition coefficient K_d, is given by adsorption isotherms which have been either measured, or calculated from the partition of individual compounds between water and octanol, K_{ow}. The binding of hydrophobic organic compounds to soil particles is due predominantly to adsorption sites on the organic carbon present in soils, identified by the partition coefficient K_{oc}.

For the more volatile compounds such as HCB and the HCHs, evaporation will be the major loss mechanism over a relatively short period of time. For compounds with higher adsorption, such as DDT, residual quantities may

[15] A. Di Guardo, R. J. Williams, P. Matthiessen, D. N. Brooke and D. Calamari, *Environ. Sci. Pollu. Res.*, 1994, **1** 151.

[16] P. Matthiessen, D. Sheahan. R. Harrison, M. Kirby, R. Rycroft, A. Turnbull, C. Volkner and R. Williams, *Ecotoxicol. Environ. Safe.*, 1995, **30**, 111.

[17] G. H. Willis and L. L. McDowell, *Environ. Toxicol. Chem.*, 1982, **1**, 267.

remain bound up within soils and stream sediments for many years. An extensive discussion of evaporative losses from different soil types under varying meteorological conditions has been presented by Taylor and Spencer.[18] It is worth noting that evaporation of pesticides from field soils has been shown to be largely dependent on sufficient moisture being present to allow an aqueous monolayer to exist around soil particles. If surface drying reduces this, evaporation rates will be reduced irrespective of increasing temperatures.[18]

The partition of compounds between water and air is given by the Henry's law constant, *H*, which may be defined as the ratio of a compound's vapour pressure to its solubility, as indicated in equation (1), or alternatively as the dimensionless equilibrium concentration in air over the corresponding concentration in water, K_{aw} (equation 2).

$$H = \frac{\text{vapour pressure}}{\text{aqueous solubility}} \tag{1}$$

$$K_{aw} = \frac{\text{air concentration}}{\text{water concentration}} \tag{2}$$

The value of *H* provides the means to describe vapour phase equilibrium between water and air. This can be used to help define vapour phase washout by rain, and also the oceanic gas phase exchange between marine air masses and associated surface water.

Atmospheric Distribution

The most important facet of organochlorine mobility following release will be vapour phase transfer to the atmosphere and subsequent deposition mechanisms. Particle transfer to the atmosphere by mechanical means may be important in some circumstances, although vapour adsorption to aerosols will be the major source of particle-bound pesticides in the atmosphere. A compound's partition between the particle and vapour phase has been identified as the critical factor in defining subsequent loss mechanisms from the atmosphere.[19] The principal parameters involved in the partitioning between the adsorbed and vapour phases in the atmosphere include ambient temperature, aerosol characteristics and the vapour pressure of individual compounds. A relationship (equation 3) for estimating this was presented by Junge in 1976[20] where the adsorbed fraction, ϕ, is calculated from aerosol surface area, θ, and the vapour pressure, *P*. The coefficient *c* will be dependent upon ambient temperature and certain aerosol characteristics, thus indicating seasonal effects, and the aerosol source will have significant impact on partitioning. It has been found that HCH and HCB exist predominantly ($>95\%$) in the vapour phase, whereas DDT has been found to be

---

[18] A. W. Taylor and W. F. Spencer, in *Pesticides in the Soil Environment*, ed. H. H. Cheng, Soil Science Society of America, 1990, pp. 213–269.

[19] T. F. Bidleman, *Environ. Sci. Technol.*, 1988, **22**, 361.

[20] C. E. Junge, in *Fate of Pollutants in the Air and Water Environments*, Wiley, New York, 1977, part 1, p. 726.

often extensively particle bound (> 50%). Toxaphene, being a complex mixture of many congeners, has not been well investigated but one recent report has found only the more volatile and less soluble congeners around Lake Baikal, Russia. The authors suggested this to be due to the longer atmospheric lifetime of insoluble vapour phase components providing evidence for their long-range transport.[21]

$$\phi = \frac{c\theta}{(P + c\theta)} \tag{3}$$

Theoretical and experimental aspects to determine these parameters, and associated deposition, have recently been extensively investigated by Pankow.[22] The principal deposition mechanisms can be divided between wet and dry processes, and particle and vapour processes. Wet deposition of particles from the air may occur due to 'washout' by falling rain, or by 'rainout' of small condensation nuclei at cloud level. Dry deposition of particles will be a combination of gravitational settling for larger particles (> 10 μm), impaction for smaller particles and diffusion for the smallest particles (< 0.1 μm). It is known that particles in the range 0.1–1 μm have the longest atmospheric lifetime.[23] Vapour phase deposition will also be a combination of wet deposition by dissolution into falling rainfall, and dry deposition onto water or terrestrial surfaces. The mechanisms of vapour and particle scavenging by rain have been considered in detail by Ligocki *et al.*[24,25] For the less hydrophobic organochlorines, such as HCH, vapour phase scavenging by rain becomes significant, whereas for the more hydrophobic DDT, particle dry deposition will be predominant.

The vapour phase adsorption to vegetation has been an area of specific interest both from the use of passive biomonitors to determine qualitative airborne levels, and also with respect to the resultant potential bioaccumulation. Uptake by biological material will be explored in more detail later.

The importance of each mechanism will depend upon the relative persistence and physical properties of individual compounds. Gas phase bi-directional fluxes allow rapid equilibrium to be achieved between air and oceanic surface water. This has been noted for the more volatile compounds, such as the HCHs, to be the controlling factor in their rapid global distribution.[26] Vapour phase scavenging by rain quickly removes airborne HCH, which will then easily evaporate in cycles which 'bounce' the chemical according to global weather patterns. This has been noted as the 'grasshopper' effect and leads to the idea of global fractionation where toxic organochlorines tend to be transported to, and accumulate in, polar regions.

An overall review of the long-range atmospheric transport of pesticides covering these issues in detail has been recently published.[27] Table 3 provides

[21] L. L. McConnell, W. E. Cotham and T. F. Bidleman, *Environ. Sci. Technol.*, 1993, **27**, 1304.

[22] J. F. Pankow, *Atmos. Environ.*, 1987, **21**, 2275.

[23] K. W. Nicholson, *Atmos. Environ.*, 1988, **22**, 2653.

[24] M. P. Ligocki, C. Leuenberger and J. F. Pankow, *Atmos. Environ.*, 1985, **19**, 1609.

[25] M. P. Ligocki, C. Leuenberger and J. F. Pankow. *Atmos. Environ.*, 1985, **19**, 1619.

[26] H. Iwata, S. Tanabe and R. Tatsukawa, *Mar. Pollut. Bull.*, 1993, **26**, 302.

[27] D. A. Kurtz, *Long Range Transport of Pesticides*, Lewis, Michigan, 1990.

Table 3 Environmentally important physical properties of the organochlorine pesticides

	Molecular weight	Melting point (°C)	Vapour pressure (mPa)[a] at 20°C	Solubility (mol m^{-3})[a] at 20°C	log K_{ow}	Henry's law (Pa m^3 mol^{-1}) at 20°C
α-HCH	291	158	73	0.084	3.8	0.87
Lindane	291	112	25	0.19	3.8	0.13
HCB	285	231	130	0.019	6.0	7.12
DDT	354.5	109	0.15	6.5×10^{-5}	6.0	2.36
Dieldrin	381	176	19	0.017	3.7	1.12
Aldrin	365	104	36	3.9×10^{-4}	3.0	91.2
Endrin	381	209	1.5	0.046	3.2	0.033
Toxaphene	414	65–90	~1	very low	–	0.42

[a]Data given as the sub-cooled vapour pressure and sub-cooled aqueous solubility.

some of the important environmental characteristics of the organochlorine pesticides considered here. It should be mentioned that published data of such physical and chemical properties can vary considerably, depending upon the source. The data in Table 3 were collated by Suntio *et al.*[28] from a broad range of previously published data.

Clearly, ultimate loss from the atmosphere will be gain for adjacent media, with potentially significant ecological effects which may be far from the sites of original use. From both a human and an ecological point of view, the most important partition will be that to biological material.

Bioaccumulation of the Organochlorine Pesticides

Partition into biological material and subsequent bioaccumulation through food chains is perhaps the most significant aspect of environmental contamination by the persistent organochlorines. This section will identify some of the mechanisms involved.

As seen earlier with respect to soil adsorption, hydrophobic contaminants preferentially partition to organic materials rather than water, air or inorganic substrates. Two major routes into the food chain can be identified as direct vapour phase adsorption into vegetation,[29] or via consumption of contaminated suspended organic detritus by small aquatic organisms.[30] Bioaccumulation can be seen to follow the complex food webs that exist in nature, with residues occurring predominantly in fatty tissues. Thus, low-level contamination of the lower species may lead to significant magnification effects in higher animals.

Similar adsorption into growing vegetation, related to the lipid content of leaves, has also been proposed,[31] where it has been shown that there is little

[28] L. R. Suntio, W. Y. Shiu, D. Mackay, J. N. Seiber and D. Glotfelty, *Rev. Environ. Contam. Toxicol.*, 1988, **103**, 1.

[29] P. Larsson, L. Okla and P. Woin, *Environ. Sci. Technol.*, 1990, **24**, 1599.

[30] V. J. Bierman, *Environ. Sci. Technol.*, 1990, **24**, 1407.

[31] D. C. G. Muir, M. D. Segstro, P. M. Welbourn, D. Toom, S. J. Eisenreich, C. R. Macdonald and D. M. Whelpale, *Environ. Sci. Technol.*, 1993, **27**, 1201.

mobility within plant leaves, stems or roots for hydrophobic chemicals with log K_{ow} values between 3 and 7.[32] A recent report highlighted the potential scale of this sink for such atmospheric pollutants.[33] This research estimated 44% of the mass budget of related polyaromatic compounds emitted over the Eastern USA would be removed into vegetation and from there into the resulting soil-bound humic materials. Studies have correlated the rate of deposition in selected Scandinavian sites directly with uptake by small mammals in which contamination was seen to reflect regional differences in atmospheric deposition.[29]

Considerable attention has been given to the investigation of organochlorine residues in the marine polar ecosystems where the two effects of global distillation and bioaccumulation combine to produce high residues in predatory species such as marine mammals, fish and sea birds. As these compounds preferentially reside in fatty tissues, the high fat content of many of the northern marine species, and the inherited contamination through lactation, indicate that such ecosystems may be some of the most vulnerable.

Sub-lethal bioaccumulation in selected species and sedimentary materials has provided a convenient source of passive 'indicators' which can be used to assess the impact of current (and past) environmental contamination. The relative combination of each of the organochlorines may then provide a qualitative and quantitative fingerprint of contamination. Pine needles have proved a useful medium by which to compare relative atmospheric levels across Europe and Africa.[34,35]

Global Distribution and Modelling

It is apparent, from observed levels in the atmosphere, that highly persistent volatile compounds will diffuse around the planet irrespective of the slow interhemispheric atmospheric transfer rates.[36] Examples of this can be seen in the relatively even distribution of the chlorofluorocarbons (CFCs) and HCB, both very volatile and stable compounds in the atmosphere. For less volatile compounds such as DDT, longer term trends suggest a decline in temperate, northern regions and a resulting shift towards more southerly latitudes, reflecting this compound's changing use patterns and also indicating slower global distribution effects of this more persistent but particle-bound compound. Other effects can be identified in the fingerprint of PCBs and toxaphenes, which contain mixtures of compounds varying from more volatile species to predominantly particle-bound species. It has been found that the relative concentration of certain congeners increases with time and distance from source regions.[37] More volatile compounds will have longer atmospheric lifetimes, but may be subject to

---

[32] E. Bacci, M. J. Cerejeira, C. Gaggi, G. Chemello, D. Calamari and M. Vighi, *Chemosphere*, 1990, **21**, 525.

[33] S. L. Simonich and R. A. Hites, *Nature*, 1994, **370**, 49.

[34] D. Calamari, E. Bacci, S. Focardi, C. Gaggi, M. Morosini and M. Vighi, *Environ. Sci. Technol.*, 1991, **25**, 1489.

[35] D. Calamari, P. Tremolada, A. Di Guardo and M. Vighi, *Environ. Sci. Technol.*, 1994, **28**, 429.

[36] S. Tanabe, R. Tatsukawa, M. Kawano and H. Hidaka, *J. Oceanogr. Soc. Jpn.*, 1982, **38**, 137.

[37] L. Barrie, T. Bidleman, D. Dougherty, P. Fellin, N. Grift, D. Muir, B. Rosenberg, G. Stern and D. Toom, *Chemosphere*, 1993, **27**, 2037.

greater decomposition. However, more particle-bound congeners will be more susceptible to removal by dry deposition. The presence of more regional distributions can be detected, for example in the ratio of the relatively less persistent γ-HCH to α-HCH. Increasing ratios of γ/α HCH seen across Europe and North America during the spring indicate fresh sources of lindane due to agricultural use.[38]

Mention has already been made of mathematical models which simulate partitioning in the environment. This has been facilitated by the introduction of fugacity principles to environmental modelling, which simplifies the linking of complex partition and rate constants in many of the current multimedia environmental models. A detailed explanation of the ideas involved, and their application, has recently been published by Mackay.[39]

Modelling the impact of organochlorines on a regional scale has been undertaken at length for the Great Lakes of North America, where mass budgets and fluxes have been calculated for many compounds.[40,41] An adventurous extension of such models has been the attempt to simulate global distribution based on a knowledge of climatic conditions in latitudinal zones, and the chemical parameters of persistent compounds.[13] This approach proved useful in identifying the global fractionation effect described earlier. A major input parameter for such simulation models will involve release statistics, for which there are often only inaccurate 'guestimates'. An alternative approach to this problem involves an assessment of residue concentrations in various environmental media from around the planet. The global distribution of organochlorine pesticides in vegetation due to vapour phase bioaccumulation from the atmosphere has been mapped by one research group.[34] It was found that HCB levels increased towards the poles. Interestingly, high-altitude equatorial samples were similar to sea level samples from higher or lower latitudes. This was attributed to the predominant influence of temperature on the bioaccumulation of this compound.

5 Environmental Occurrence

A comparison of published levels of organochlorines from around the planet gives a good indication of the typical contamination that might be expected in different media. Clearly, much of these data have been collected with specific objectives in mind, which may have influenced the results obtained. In addition, it should be remembered that such comparisons will always contain an unquantifiable degree of sampling and analytical bias. Some of the data in the following tables reflect authors who have themselves collated a number of different sources. The overall implication of these data is that their release over the last 50 years has generally exceeded the rate at which degradation mechanisms can remove them from the environment.

[38] M. Oehme, J. Haugen and M. Schlabach, *Sci. Total Environ.*, 1995, **116**, 139.
[39] D. Mackay, *Multimedia Environmental Models – The Fugacity Approach*, Lewis, 1991.
[40] S. J. Eisenreich, B. B. Looney and J. D. Thornton, *Environ. Sci. Technol.*, 1981, **15**, 30.
[41] R. M. Hoff, T. F. Bidleman and S. J. Eisenreich, *Chemosphere*, 1993, **27**, 2047.

Organochlorine Pesticides in the Atmosphere

Table 4 provides data on pesticides in air from remote, marine, rural and urban areas. The continuing use of HCH and DDT is clearly seen in the high concentrations around the Bay of Bengal, as for toxaphene and DDT in the older data from the USA (1979). Data for HCB show a remarkable uniformity, indicative of the global diffusion of this compound. Arctic levels are generally lower, although notable exceptions are the more volatile HCH and HCB molecules. The α/γ HCH ratio reported for the Arctic is considerably above that found in technical HCH, which is indicative of considerable isomeric conversion of both lindane and the technical mixture. A low α/γ HCH ratio can be seen to represent areas of continuing lindane use, such as in the UK and Great Lakes regions.

Occurrence in Rain

A selection of rain concentrations derived from the literature is given in Table 5. The current use of lindane is again evident in the isomeric ratios detected in Europe compared to other regions. Within the UK, current levels of DDT, DDE and dieldrin are significantly lower than when first measured in the late 1960s. Other than the Thames Valley data, HCB concentrations are uniformly low, reflecting this compound's low aqueous solubility and its occurrence predominantly in the vapour phase. The relatively high levels of DDT in rain are due probably to greater particle phase partitioning and subsequent scavenging by rain or inclusion in rain data due to dry deposition into collectors.

Contamination of the Oceans

Less information exists with respect to contamination of surface seawater, although it is acknowledged that the gas-phase exchange of the semi-volatile organochlorines from the atmosphere will probably influence their occurrence more than direct riverine discharges. Recent reports suggest that global distribution will be largely controlled by gas-phase equilibrium between the oceans and the atmosphere.[42] A logical extension of this is the probability that the oceans will become a diffuse source of such compounds for many years as increasing restriction reduces their release. Recent reports on α-HCH levels in the atmosphere have indicated a potential decline such that oceans are already a net source of this compound in the northern hemisphere.[43] Data from the Great Lakes and Lake Baikal reveal generally greater contamination, especially for the DDT compounds, due probably to the additional influence of adsorption onto eroded suspended sediments. A selection of results from oceans around the planet is given in Table 6.

[42] J. Schreitmuller and K. Ballschmiter, *Environ. Sci. Technol.*, 1995, **29**, 207.

[43] T. F. Bidleman, L. M. Jantunen, R. L. Falconer, L. A. Barrie and P. Fellin, *Geophys. Res. Lett.*, 1995, **22**, 219.

Table 4 Organochlorine concentrations in air (pg m^{-3})

	Chlordanes	α-HCH	Lindane	HCB	Toxaphene	Dieldrin	DDE	DDT	Date
Southern England	–	39	408	43	–	40	14	17	1992–93[b]
Thames Valley, UK	–	29	140	25	–	14	–	29	1987–90[c]
Southern Sweden	8.4		489[a]	64	25	–		7.2[a]	1983–85[d]
Texas	1050	710	230	210	1770	80	260	60	1979–80[e]
Great Lakes	–	300	2000	200	–	50		30[a]	1978[f]
Bermuda	22	122	12	89	–	20	–	–	1983–84[g]
Southern Ontario	14	145	60	54	–	46	53	20	1988–89[h]
Bay of Bengal	24	8600	1100	–	–	–	19	240	1989–90[i]
South China Sea	39	810	500	–	–	–	17	40	1989–90[i]
Mediterranean Sea	15	150	75	133	–	–	10	4	1989–90[i]
Norwegian Arctic	1.4	503	33	130	–	–	–	–	1982–84[j]
Arctic	3.8	450	33		31	1.6	–	1.2	1980s[k]
North Atlantic	14	260	53	126	–	13	6	6	1980s[l]

[a] Sum of DDT + metabolites, or sum of $\alpha + \gamma$HCH.
[b] A. B. Turnbull, PhD Thesis, Birmingham University, 1995.
[c] I. T. Cousins and C. D. Watts, *Atmospheric Sources of Pollution: Inputs of Trace Organics to Surface Waters*, National Rivers Authority R & D Report 20, NRA, Bristol, 1995.
[d] T. F. Bidleman, U. Wideqvist, B. Jansson and R. Soderlund, *Atmos. Environ.*, 1987, **21**, 641.
[e] E. Atlas and C. S. Giam, *Science*, 1981, **211**, 163.
[f] S. J. Eisenreich, B. B. Looney and J. D. Thornton, *Environ. Sci. Technol.*, 1981, **15**, 30.
[g] A. H. Knap, K. S. Binkley and R. S. Artz, *Atmos. Environ.*, 1988, **22**, 1411.
[h] R. M. Hoff, D. C. G. Muir and N. P. Grift, *Environ. Sci. Technol.*, 1992, **26**, 266.
[i] H. Iwata, S. Tanabe, N. Sakai and R. Tatsukawa, *Environ. Sci. Technol.*, 1993, **27**, 1080.
[j] J. M. Pacyna and M. Oehme, *Atmos. Environ.*, 1988, **22**, 243.
[k] W. E. Cotham and T. F. Bidleman, *Chemosphere*, 1991, **22**, 165.
[l] GESAMP, *The Atmospheric Input of Trace Species to the World Ocean*, Joint Group of Experts on the Scientific Aspects of Marine Pollution (GESAMP), Report and Studies No. 38, World Meteorological Organisation, 1989.

Table 5 Organochlorine concentrations in rain (ng l^{-1})

	Chlordanes	α-HCH	Lindane	HCB	Toxaphene	Dieldrin	DDE	DDT	Date
Rural England	—	1.9	52	0.5	—	1.1	0.5	1.1	1992–93[a]
Thames Valley, UK	—	5.1	12	8.1	—	2.2	—	14	1987–90[b]
Northern Italy	—	—	—	—	—	—	5.9	13	1985–88[c]
Monaco	—	6.1	31	0.7	25	—	1.5	3.4	1985[d]
Texas	2.1	12	3.2	0.4	22	0.8	1.0	2.1	1979–80[e]
Great Lakes	—	12	4.6	0.055	—	0.45	0.085	0.46	1981–84[f]
Bermuda	0.08	0.85	0.012	0.02	—	0.16	—	0.03	1980s[g]
UK	—	—	60	—	—	7.6	19	46	1967[h]

[a] A. B. Turnbull, PhD Thesis, Birmingham University, 1995.
[b] I. T. Cousins and C. D. Watts, *Atmospheric Sources of Pollution: inputs of trace organics to surface waters*, National Rivers Authority **R & D** Report 20, NRA, Bristol, 1995.
[c] S. Galassi, E. Gosso and G. Tarari, *Chemosphere*, 1993, **27**, 2287.
[d] J. Villeneuve and C. Cattini, *Chemosphere*, 1986, **15**, 115.
[e] E. Atlas and C. S. Giam, *Science*, 1981, **211**, 163.
[f] W. M. J. Strachan, *Environ. Toxicol. Chem.*, 1985, **4**, 677.
[g] GESAMP, *The atmospheric input of trace species to the world ocean*, Joint Group of Experts on the Scientific Aspects of Marine Pollution (GESAMP), Report & Studies No. 38, World Meteorological Organisation, 1989.
[h] K. R. Tarrant and J. O. G. Tatton, *Nature*, 1968, **219**, 725.

Table 6 Organochlorine concentrations in surface sea and lake water (pg l^{-1})

	Chlordanes	α-HCH	Lindane	HCB	Toxaphene	Dieldrin	DDE	DDT	Date
Mediterranean	3.7	180	150	–	–	–	1.2	1.3	1989–90[b]
South China Sea	9.3	380	97	–	–	–	1	6	1989–90[b]
Arctic	3.7	5970	824	19	380	15		2[a]	1980s[c]
Bay of Bengal	7.9	610	110	–	–	–	1.4	8.6	1989–90[b]
Southern Ocean	3.4	28	8.2	–	–	–	0.5	0.5	1989–90[b]
Bering Sea	3.4	1500	190	–	–	–	0.9	0.1	1989–90[b]
Lake Baikal	46.4	1100	240	20	64	–		87[a]	1991[d]
Great Lakes	36–140	1300	420	18–73	–	–	18–45	–	1980s[d]

[a] sum of DDT + metabolites.
[b] H. Iwata, S. Tanabe, N. Sakal and R. Tatsukawa, *Environ. Sci. Technol.*, 1993, **27**, 1080.
[c] W. E. Cotham and T. F. Bidleman, *Chemosphere*, 1991, **22**, 165.
[d] J. R. Kucklick, T. F. Bidleman, L. L. McConnel, M. D. Walla and G. P. Ivanov, *Environ. Sci. Technol.*, 1994, **28**, 31.

Seasonal Effects

Seasonal effects have been detected and must clearly reflect the agricultural requirements within regions of use. Springtime transport of lindane to the Arctic was mentioned earlier and can be related to increased use in Northern Europe during this period.[38] This is also reflected in the seasonal trend of α/γ HCH ratios in both air and rain over the UK.[44] However, meteorological factors will also considerably influence this as Arctic air masses are relatively static during the winter period, with increased movement from lower latitudes during the spring and summer. Temperature has also been implicated in the seasonal occurrence and long-range transport of semi-volatile persistent compounds in the atmosphere due to the relative evaporation rates and vapour–particle partitioning so induced.[45]

Bioaccumulation in Biological Tissues

A large amount of data has been published on residue levels in many different species. Much of this data will be specific to local and regional circumstances, especially with respect to higher organisms, and is often most relevant to health and toxicity issues. However, Table 7 provides data on levels recorded in vegetation, considered to be more related to ambient atmospheric levels. The data in Table 7 provide a reasonable indication of the local use of lindane in the UK, DDT in Czechoslovakia and also of the low level contamination of Antarctica due to long-range transport.

Marine mammals have also been assessed for organochlorines due to their position at the top of the oceanic food chains. Feeding habits, location and metabolic activity of individual species have all been shown to influence residue levels. A review of residue levels in seals around the Japanese coast found decreasing DDT:DDE ratios in fat since 1971, with relatively stable DDE levels of around 5–$15\,\mu\mathrm{g\,g}^{-1}$.[46] Stranded whales have also provided an opportunity to assess bioaccumulation into the large and mobile fat compartments which these creatures carry. Not surprisingly, contamination varies considerably according to species, with humpbacks and minkes containing between 0.1–7 and 0.1–23 μg g^{-1} of DDE, respectively, in their fat, whereas for killer whales values between 0.1–250 μg g^{-1} have been recorded.[47] Another recent study on otters in southwest England found mean levels of 10.4 and 22.3 μg g^{-1} in body fat for dieldrin and DDE, respectively.[48] Levels of the HCH isomers, although often recorded, are generally much lower due to the more common occurrence of metabolic degradation mechanisms.

In the light of the growing restrictions on organochlorine use, the temporal changes in commonly sampled media such as human milk and adipose tissue provide data on the efficacy of control legislation. A Dutch study has assessed

[44] A. B. Turnbull, PhD Thesis, Birmingham University, 1995.

[45] R. M. Hoff, D. C. G. Muir and N. P. Grift, *Environ. Sci. Technol.*, 1992, **26**, 266.

[46] S. Tanabe, J. K. Sung, D. Y. Choi, N. Baba, M. Kiyota, K. Yoshida and R. Tatsukawa, *Environ. Pollut.*, 1994, **85**, 305.

[47] U. Varanasi, J. E. Stein, K. L. Tilbury, J. P. Meador, C. A. Sloan, R. C. Clark and S. L. Chan, *Sci. Total Environ.*, 1994, **145**, 29.

[48] C. F. Mason and S. M. Macdonald, *Sci. Total Environ.*, 1994, **144**, 305.

Table 7 Organochlorine concentrations determined in vegetation (ng g^{-1} dry weight)

	α-HCH	Lindane	HCB	DDE	DDT	Date
Antarctic: lichens	0.3	0.7	0.5	0.2	nd	1985–88[a]
Northern Sweden	3.3	12.4	0.25	0.2	1.8	1975[b]
Bavaria: pine needles	8.9	13.5	2.5	1.8	5.4	1986[c]
UK: pine needles	4.0	37	4.2	1.3	1.8	1992–93[d]
Czech Rep: pine needles	14	27	13	24	10	1991–92[e]
W. Africa: mango leaves	0.9	1.5	<0.1	3.3	21	1985–87[f]

[a]D. Calamari, E. Bacci, S. Focardi, C. Gaggi, M. Morosini and M. Vighi, *Environ. Sci. Technol.*, 1991, **25**, 1489.
[b]J. P. Villeneuve, E. Holm and C. Cattini, *Chemosphere*, 1985, **14**, 1651.
[c]A. Reischl, M. Reissinger and O. Hutzinger, *Chemosphere*, 1987, **16**, 2647.
[d]A. B. Turnbull, PhD Thesis, Birmingham University, 1995.
[e]D. Calamari, P. Tremolada, A. Di Guardo and M. Vighi, *Environ. Sci. Technol.*, 1994, **28**, 429.
[f]P. Tremolada, D. Calamari, C. Gaggi and E. Bacci, *Chemosphere*, 1993, **27**, 2235.

contamination of human adipose tissue between the years 1968–1986. This research found a dramatic decrease in DDT levels from 1.5 to around $0.2\ \mu g\ g^{-1}$ (fat basis). However, a similar decrease in the more predominant DDE levels was not evident, with levels remaining around $2–3\ \mu g\ g^{-1}$.[49] The decreasing ratio of DDT:DDE reflects the greater stability of DDE in the environment, and the restriction of DDT use in the developed world. A study of Welsh citizens in 1991 revealed DDE in fat of $1.4\ \mu g\ g^{-1}$, and also noted little decrease in this compound over recent decades.[50] The relatively stable DDE levels demonstrates the conservative behaviour of this compound, to some extent passing through generations through lactation, although contamination appears to increase with the age of subjects. Levels of DDE in human milk in $\mu g\ g^{-1}$ fat have been found to be relatively stable in Europe at around $0.5\ \mu g\ g^{-1}$, except for the low level reported by the UK government. Residues of HCB and PCBs also occur and appear stable in these human studies, Table 8. Levels of the HCH isomers are relatively low (irrespective of the continuing use of lindane), predominantly due to the more efficient metabolic removal mechanisms mentioned above.

6 Environmental Toxicity

The overall importance of research into pesticides in the environment is to understand and limit their potential toxic effects beyond target organisms, and to ensure resistance in these organisms is prevented. The post-war experiment with organochlorines rapidly produced acute toxic effects in many non-target species and, additionally, many incidences of resistance occurred. Initially there was incredulity to claims of pesticides killing predatory species across Europe and America, although with hindsight their environmental persistence, mobility and bioaccumulation provided clear and logical reasons for these effects. In addition, the impact of combinations of these pesticides along with other persistent pollutants such as PCBs, PAH and dioxins may produce complex synergistic effects which, where relevant, may take many years to elucidate.

Acute Toxic Effects

Much of the evidence for acute toxicity was derived from the 1950s and 1960s when the direct contamination of specific food chains was seen. High levels could be observed in earthworms following field use, leading directly to the contamination of many bird species. In addition, seed-eating birds were consuming large quantities of organochlorine seed dressings when feeding on treated seed. The contamination rapidly moved and became magnified in predatory species. As a result, large numbers of bird deaths were recorded, some even falling from the air in mid-flight. A considerable number of these victims, when assessed for organochlorine contamination of selected body tissues, were found to contain potentially lethal levels.[4]

The species most commonly reported dead in the UK were the sparrowhawk, kestrel, tawny owl and barn owl, whose bodies were sent to Monks Wood

[49] P. A. Greve and P. Van Zoonen, *Int. J. Environ. Anal. Chem.*, 1990, **38**, 265.
[50] R. Duarte-Davidson, S. C. Wilson and K. C. Jones, *Environ. Pollut.*, 1994, **84**, 79.

Table 8 Organochlorine concentrations determined in human milk (μg g^{-1} of fat)

	α-HCH	Lindane	HCB	DDE	DDT	Date
UK[a]	0.08[b]	–	–	0.56	0.36	1963–64[c]
UK[a]	0.02[b]		0.008	0.07	<0.008	1989–90[c]
Wales	–	–	0.13	0.45	0.06	1990–91[d]
Norway	–	–		0.50	–	1979–85[e]
Holland	<0.01	<0.01	0.19	0.82	0.04	1983[f]
Sweden	–	–	0.05	0.05	0.05	1989[g]

[a]Calculated from total milk data assuming a 2.5% fat content.
[b]Sum of HCH isomers.
[c]HMSO, *Report of the Working Party on Pesticide Residues* 1988–90, Food Surveillance Paper No. 34, HMSO, London, 1992.
[d]R. Duarte-Davidson, S. C. Wilson and K. C. Jones, *Environ. Pollut.*, 1994, **84**, 79.
[e]J. U. Skaare and A. Polder, *Arch. Environ. Contam. Toxicol.*, 1990, **19**, 640.
[f]P. A. Greve and P. Van Zoonen, *Int. J. Environ. Anal. Chem.*, 1990, **38**, 265.
[g]K. Noren, *Sci. Total Environ.*, 1993, **139/140**, 347.

Experimental Station in the early 1960s. Experiments showed that small mammals feeding on seed dressed with dieldrin, when fed to kestrels could cause death within seven days. Additionally, it was noted that the abnormal behaviour of contaminated voles would probably draw the attention of predatory species. Among other animals also implicated with acute toxic effects during this period were foxes, badgers, otters and bats. Complex metabolic reactions within different species and their particular feeding habits apparently led to the divergent responses to the organochlorine threat. Species which seasonally store and then utilize fat reserves, such as hibernating or migrating animals, are suspected of mobilizing lethal doses of organochlorines from their own bodies.

Most of the reported deaths during this period were associated with either dieldrin or DDT, with HCH being less persistent in biological tissue, and generally accumulating at lower levels. Table 9 gives the relative toxicity of the different organochlorines with respect to rats and fish.

Chronic and Synergistic Effects

Two other birds were also affected in the early 1960s which perhaps drew attention to the encroaching problem more than all the other species. These were the golden eagle and peregrine falcon. The UK government commissioned surveys which revealed a serious decline in their populations. The peregrine falcon dropped to 40% of its pre-war population. It was reported that there were less breeding pairs and an increase in eggshell breakage was observed for the golden eagle. Such sub-lethal effects on reproduction can potentially be as devastating to a species as outright mortality; it just takes a little longer. The cause of eggshell thinning in most species of birds is now confirmed as due to contamination by DDT and its metabolites, which affect calcium metabolism.[51,52]

Many longer-term sub-lethal effects are now becoming more recognized Environmental estrogens (molecules which mimic the female hormone) have now been identified as significant factors in the development of reproductive abnormalities in many species.[53] Higher concentrations of non-chlorinated estrogenic compounds such as phthalate esters and nonylphenyls are implicated with hermaphrodicity in fish, whereas much lower levels of DDE are implicated with the development of cancers in human breast cell experiments.[54] It is reported that where 10 compounds at low levels individually show no estrongenic effects, combined they have a positive influence on the growth of cancerous cells.[55]

Potential Human Toxicity

The cause of much current concern over organochlorines has been the evidence of human bioaccumulation and the possibility of significant health effects. Acceptable

[51] W. H. Stickel, in *Ecological Toxicological Research*, ed. A. D. McIntyre and C. F. Mills, Plenum Press, New York, 1975, pp. 25–74.

[52] D. B. Peakall and J. Lincer, *Environ. Pollut.*, 1996, **91**, 127.

[53] T. Colborn, F. S. V. Saal and A. M. Soto, *Environ. Health Perspect.*, 1993, **101**, sup. 5, 378.

[54] M. S. Wolff and P. G. Toniolo, *Environ. Health Perspect.*, 1995, **103**, sup. 7, 141.

[55] A. M. Soto, K. L. Chung and C. Sonnenschein, *Environ. Health Perspect.*, 1994, **102**, sup. 4, 380.

Table 9 Relative toxicity of some organochlorine pesticides[a]

		Units	*HCB*	*DDT*	*Lindane*	*Dieldrin*
Rat	LD_{50}	mg kg^{-1}	10 000	113–118	88–270	46
Fish[b]	LC_{50}	mg l^{-1}	0.05–0.2	Toxic to fish	160	0.02–0.09

[a]Data from the *The pesticide manual*, British Crop Protection Council, Croydon, 1991.
[b]Accurate LC_{50} data will be species specific and depend upon the period of exposure.

Daily Intake values (ADIs) for the various compounds have been developed and are generally much lower than dietary inputs.[7] Specific concern with respect to infants during lactation has been investigated by the World Health Organisation, who felt the benefits of breast feeding outweighed the potential dangers of organochlorine contamination.[56] Recent empirical reports on the possible impact of organochlorines on the increasing infertility of men, and the potential increased risk of breast cancer in women, is also clearly of considerable concern. A study on infertile men found increased levels of DDT metabolites in patients' blood.[57] Similar statistical work on women in New York found a significant increased risk of breast cancer in those with higher DDE levels in their adipose tissue.[58] It has also been suggested that such estrogenic compounds may be affecting testicular cancer.[1] Although some reports on environmental estrogens and breast cancer prove inconclusive,[59] it is generally accepted that the issue warrants further investigation.

7 Conclusions

Organochlorine pesticides remain present in the wider environment at low levels. Their physical properties have allowed them to accumulate in fatty biological tissues, where they may remain over decades unless released by lactation or predation. Initial acute ecological effects have led to increasing regulation throughout the world, and many of the threatened species such as the birds of prey are now seen to be recovering in the UK. Owing to concentration effects through food chains, and a tendency for these compounds to migrate to polar regions, northern marine mammals and birds are considered some of the most vulnerable species.

Current concerns with estrogenic compounds and their influence on a broad range of reproductive disorders, along with possible carcinogenic effects, have renewed interest in chlorinated pesticides. The need for ongoing scientific research to assess the potential risk posed by these compounds seems inevitable.

[56] World Health Organisation, *PCBs, PCDDs and PCDFs in Breast Milk: Assessment of Health Risks*, WHO, Geneva, 1988.
[57] A. Pines, S. Cusoc, P. Ever-Hadani and M. Ron, *Arch. Environ. Contam. Toxicol.*, 1987, **16**, 587.
[58] M. S. Wolff, P. G. Toniolo, E. W. Lee, M. Rivera and N. Dublin, *J. Nat. Cancer Inst.*, 1993, **85**, 648.
[59] T. Key and G. Reeves, *Br. Med. J.*, 1994, **308**, 1520.

Studies of Polychlorinated Biphenyls in the Great Lakes

DEBORAH L. SWACKHAMER

1 Introduction

Polychlorinated biphenyls (PCBs) have proved to be a blessing and a curse to the modern world. Invented in the late 1800s in Germany, they were commercially manufactured by Monsanto in the US beginning in 1929 under the trade name Aroclor.[1] Other countries, including Great Britain, France, Japan, Germany, Italy and the former USSR, manufactured PCBs under a variety of trade names. These chemical formulations were of low reactivity, low degradability and had high dielectric constants and low vapour pressures and water solubilities. These properties made them ideally suited for an enormous number of commercial and industrial applications throughout the world. The same properties made them highly persistent in the environment, and toxic to plants, wildlife and humans.

PCBs consist of those structures having from 1 to 10 chlorines bound to a biphenyl molecule. Of the 209 possible congeners, only approximately 130 of the structures are energetically favoured in the manufacturing process. The commercial formulations, such as Monsanto's 8 Aroclor mixtures, had varying percentages of chlorine which allowed them to be tailored to a wide range of uses. More than half of all PCB uses were in electrical applications, as dielectric fluids in transformers and capacitors. They were also used in hydraulic fluids, cutting fluids, pigment carriers in paints, carbonless copy papers, adhesives, rubber formulations, surface coatings, wax extenders, carriers in pesticide formulations, fire retardants and other miscellaneous products. Their production peaked in the early 1970s, with total global production estimated at 1 200 000 tons.[2] In the US, the use and manufacture of PCBs for non-electrical applications was restricted in 1972, followed by a complete ban in 1976. Canadian, European and Japanese production stopped at about the same time.

Of the total amount of PCBs produced globally, it is estimated that 65% are still in service or landfilled, 4% have been destroyed and 31% are mobile in the

---

[1] National Academy of Science, *Polychlorinated Biphenyls*, National Research Council, Washington, 1979.
[2] S. Tanabe, *Environ. Pollut.*, 1988, **50**, 5.

environment.[2] The resistance to thermal, biological or chemical degradation means that PCBs in the environment are subject to long-range transport and long residence times. PCBs are now distributed globally, are found in the remote regions of the Arctic and Antarctic, and have established a baseline concentration in wildlife and humans. They are one of the most ubiquitous and most often occurring anthropogenic contaminants in the world.

PCB exposure can cause a wide array of toxic responses.[3,4] In plants, PCBs can interfere with the photosynthesis process.[5] Acute exposure can cause chloracne in humans. Animal studies have shown PCBs to be teratogens, carcinogens (promoters) and endocrine disrupters. Certain structures with no or a single *ortho*-chlorine have coplanar configurations, and are considered 'dioxin-like' in their toxicity. These structures bind to the TCDD or 'dioxin' receptor, and activate the aryl hydrocarbon hydroxylase (AHH) enzyme systems which in turn cause a number of effects. Some of the non-planar structures have been shown to bind to the dopamine receptor in the brain, stimulating other responses.[6] Observed effects from chronic PCB exposure include reproductive failure, deformities, behavioural and physical developmental deficiencies, wasting, liver disease and immune system impairment. A long-term study of children born to mothers who consumed large amounts of PCB-tainted fish has reported a correlation of interuterine exposure to low birth weight and smaller head circumference at birth, and developmental deficits at several time points during childhood.[7-11]

In the Great Lakes region of North America, several studies have documented the relationship between harmful effects in natural populations of wildlife and ambient PCB concentrations. Reproductive failure in mink, terns, cormorants and gulls and bird deformities have been linked to PCB exposure.[12-14] The human study described above was also conducted in the Great Lakes region. The Great Lakes have been a focal point of PCB research because of the high levels of PCBs in water and biota and their tremendous value as a natural resource, particularly the sport fishery. The lessons learned in the Great Lakes have proved

[3] S. Safe, *Mutat. Res.*, 1989, **220**, 31.

[4] R. W. Flint and J. Vena, *Human Health Risks from Chemical Exposure: the Great Lakes Ecosystem*, Lewis, Chelsea, MI, 1991. p. 27.

[5] A. Sodergren and C. Gelin, *Bull. Environ. Contam. Toxicol.*, 1983, **30**, 191.

[6] R. F. Seegal, B. Bush and K. O. Brosch, *Neurotoxicology*, 1991, **12**, 55.

[7] S. Jacobson, J. Jacobson, P. Schwartz and G. Fein, *PCBs: Human and Environmental Hazards*, ed. F. M. D'Itri and M. A. Kamrin, Butterworth, Boston, MA, 1983, p. 311.

[8] S. W. Jacobson, G. G. Fein, J. L. Jacobson, P. M. Schwartz and J. K. Dowler, *Child Develop.*, 1985, **56**, 853.

[9] J. L. Jacobson and S. W. Jacobson, in *Toxic Contaminants and Ecosystem Health: a Great Lakes Focus*, ed. M. Evans, Wiley, New York, 1988, p. 373.

[10] J. L. Jacobson, S. W. Jacobson and H. E. B. Humphrey, *J. Pediatr.*, 1989, **116**, 38.

[11] J. L. Jacobson, S. W. Jacobson, R. J. Padgett, G. A. Brumitt and R. L. Billings, *Develop. Psychol.*, 1992. **2892**, 297.

[12] T. J. Kubiak, H. J. Harris, L. M. Smith, T. R. Schwartz, D. L. Stalling, J. A. Trick, L. Silea, D. E. Docherty and T. C. Erdman, *Arch. Environ. Contam. Toxicol.*, 1989, **18**, 706.

[13] J. P. Giesy, J. P. Ludwig and D. E. Tillett, *Environ. Sci. Technol.*, 1994, **28**, 128.

[14] M. Gilbertson, in *Toxic Contaminants and Ecosystem Health: A Great Lakes Focus*, ed. M. Evans, Wiley, New York, 1988, p. 133.

valuable for understanding the dynamics of PCB contamination in other parts of the world.

The Great Lakes form part of the boundary between the United States and Canada, and consist of five major lakes with connecting channels between them (see Figure 1, for example). They are the largest body of freshwater in the world, containing more than 10% of the Earth's surface freshwater. More than 33 million Americans and Canadians reside in the Great Lakes basin, which supports multi-billion dollar shipping, manufacturing, mining and fishing industries. The long hydraulic residence times of the lakes (190 years for Lake Superior to 3 years for Lake Erie), large surface area ($244\,000\,km^3$ total) and multiple sources of PCBs make them particularly vulnerable to PCB contamination. The long foodchains and the propensity for PCBs to biomagnify result in unacceptable elevated levels of PCBs in commercial and sport fish. PCBs enter Lake Superior from the atmosphere almost exclusively;[15] the other lakes have had a mixture of atmospheric and non-atmospheric loadings.[16-18] The US and Canada have monitored PCB contamination in the Great Lakes through a variety of programmes over the past two decades. Researchers have used the lakes to understand the physical, chemical and biological processes that control the transport, behaviour and fate of PCBs in aquatic systems, and to develop models to help predict the behaviour of PCBs and other contaminants. The time trends revealed by these monitoring efforts and the current understanding of the dynamics of PCB behaviour in the Great Lakes will be discussed in detail below. Only total PCBs will be discussed; the importance of individual congeners and homologues is well-recognized and the reader is referred to the individual citations for such information.

2 PCB Concentrations in the Great Lakes

Water Concentrations and Time Trends

There have been no long-term monitoring programmes of water concentrations for the Great Lakes. The Inland Waters Directorate of the Ontario Region of Environment Canada collected surface waters from open waters of each of the lakes from 1986 to 1990, and the Great Lakes National Program Office (GLNPO) of the US Environmental Protection Agency (USEPA) collected surface waters from each of the lakes in 1993 and 1994. Isolated studies of PCBs in water have also been conducted, usually consisting of intensive studies of a given region over a narrow time-frame.[19] The data are summarized in Table 1. Detection limits and analytical difficulties precluded the collection of a more extensive database prior to the mid-1980s. Generally, concentrations of PCBs are lowest in Lake Superior and Lake Huron (currently $0.1-0.2\,ng\,l^{-1}$) because of

15 W. M. J. Strachan and S. J. Eisenreich, *Mass Balancing of Toxic Chemicals in the Great Lakes: The Role of Atmospheric Deposition*, International Joint Commission, Windsor, Ontario, Canada, 1988.

16 D. L. Swackhamer and D. E. Armstrong, *Environ. Sci. Technol.*, 1986, **20**, 879.

17 D. Mackay, *J. Great Lakes Res.*, 1989, **15**, 283.

18 S. J. Eisenreich and W. M. J. Strachan, *Estimating Atmospheric Deposition of Toxic Substances to the Great Lakes*, Great Lakes Protection Fund and Environment Canada, Burlington, Ontario, 1992.

19 D. L. Swackhamer and D. E. Armstrong, *J. Great Lakes Res.*, 1987, **13**, 24.

Figure 1 PCB concentrations (in ng m^{-3}) in the vapour phase of air over the Great Lakes measured in fall 1991 and spring 1992. Small letters indicate cruise tracks. (Adapted from SOLEC[22,69])

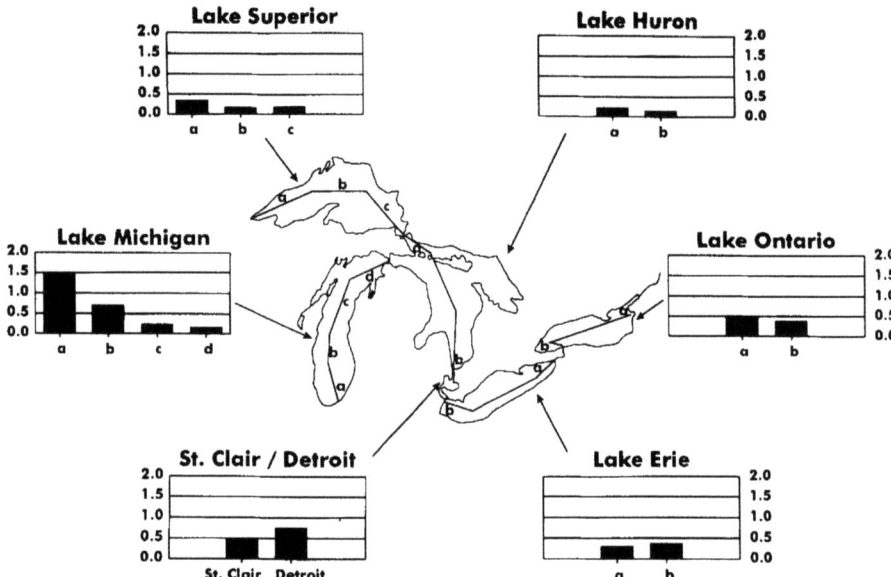

their large volumes and low population and industrialization, and higher in the other lakes (0.3–0.6 ng l^{-1}) due to the influence of their large urban and industrial centers. Available data indicate that concentrations have decreased significantly over time, following first-order kinetics.[20,21] Half-lives of PCBs in water are reported to be 3.5 years and 7.8 years in lakes Superior and Michigan, respectively. Estimates of PCB half-lives in the other lakes using the tabulated data range from 2.5 to 4.4 years. Lake Michigan has a longer half-life than the other lakes because of the influence of the large reservoir of PCBs in bottom sediments.

Concentrations near historical sources or in-place sediment sources are more elevated than open waters. These include most of the major harbours, and in particular Saginaw Bay in Lake Huron, Green Bay in Lake Michigan, Waukegan Harbour in Lake Michigan and Hamilton Harbour in Lake Ontario. Concentrations in these areas can be in the range of 20–50 ng l^{-1}.

The USEPA Great Lakes Water Quality Guidance criteria for PCBs was set at 0.017 ng L^{-1} in 1995.[22] Current data indicate that this criteria is exceeded in all of the lakes, including Lake Superior. The Canadian guideline is 1.0 ng l^{-1}, and is not exceeded in the open waters of any of the lakes at this time. The strict criteria recently established by the US and the implications for compliance are a topic of current debate.

The distribution of PCBs between the dissolved and particulate phases is

[20] J. D. Jeremiason, K. C. Hornbuckle and S. J. Eisenreich, *Environ. Sci. Technol.*, 1994, **28**, 903.

[21] R. F. Pearson, K. C. Hornbuckle, S. J. Eisenreich and D. L. Swackhamer, *Environ. Sci. Technol.*, 1996, **30**, 1429.

[22] D. S. De Vault III, P. Bertram, D. M. Whittle and S. Rang, *State of the Lakes Ecosystem Conference (SOLEC): Toxic Contaminants in the Great Lakes*, USEPA and Environment Canada, September 1995.

Table 1 Current PCB concentrations in Great Lakes open waters, and selected historical concentrations

	Current [PCB] (ng l^{-1}) (yr)	$t_{1/2}$ (yr)	*Historic [PCB]* (ng l^{-1}) (yr)
L. Superior	0.18 (1992)[a]	3.5[a]	1.7 (1978)[a]
	0.11 (1992)[b]		0.34 (1986)[c]
L. Michigan	0.46 (1995)[d]	7.8[e]	1.2 (1980)[f]
	0.42 (1992)[b]		
L. Huron	0.12 (1992)[b]	2.5[g]	0.62 (1986)[c]
L. Erie	0.44 (1992)[b]	3.6[g]	1.4 (1986)[c]
L. Ontario	0.33 (1992)[b]	2.9–4.4[g]	1.4 (1986)[c]
	0.64 (1991)[h]		

[a]Ref. 20.
[b]Ref. 66.
[c]Ref. 67.
[d]Ref. 44.
[e]Ref. 21.
[f]Ref. 19.
[g]Calculated with data shown.
[h]Ref. 24.

dependent on the concentration of suspended particulate matter, dissolved and particulate organic carbon concentrations, and the extent to which the system is at equilibrium. In general, 20–50% of the PCBs are found associated with particulates.[16,20,21] Studies of the distribution of PCBs between the dissolved and particulate phases consistently indicate that PCBs in the water column are not at equilibrium.[21,23,24]

Air Concentrations and Time Trends

The US and Canada have conducted a joint monitoring programme for contaminants in air since 1987, known as the Integrated Atmospheric Deposition Network (IADN). This database comprises the vast majority of air and precipitation concentrations available for the Great Lakes region. Air and precipitation samples from master stations around the Great Lakes shore are collected every two weeks. These data have been interpreted for seasonal and spatial trends in a number of reports.[15,18,25,26] Research by Eisenreich and colleagues has provided a more detailed picture of atmospheric PCB concentrations and dynamics for lakes Superior and Michigan[27-29] (also see below).

[23] J. E. Baker, S. J. Eisenreich and D. L. Swackhamer, in *Organic Substances and Sediments in Water*, ed. R. A. Baker, Lewis, Chelsea, MI, 1991, p. 79.

[24] D. L. Swackhamer, University of Minnesota, Minneapolis, MN, unpublished data.

[25] R. M. Hoff, D. C. G. Muir and N. P. Grift, *Environ. Sci. Technol.*, 1992, **26**, 266.

[26] R. M. Hoff, D. C. G. Muir and N. P. Grift, *Environ. Sci. Technol.*, 1992, **26**, 276.

[27] K. C. Hornbuckle, J. D. Jeremiason, C. W. Sweet and S. J. Eisenreich, *Environ. Sci. Technol.*, 1994, **28**, 1491.

[28] K. C. Hornbuckle, C. W. Sweet, R. F. Pearson, D. L. Swackhamer and S. J. Eisenreich, *Environ. Sci. Technol.*, 1995, **29**, 869.

[29] D. A. Achman, K. C. Hornbuckle and S. J. Eisenreich, *Environ. Sci. Technol.*, 1993, **27**, 75.

PCBs are mostly in the vapour phase, and are generally highest in the summer and lowest in the winter.[25,27] The most recent concentrations range from 30 to 400 pg m^{-3}. Air vapour concentrations taken over water of all five lakes from the R/V *Lake Guardian* in fall 1991 and spring 1992 are shown in Figure 1. Comparison of concentrations taken during the same season across the lakes indicates that there is little geographical variation, and that concentrations are similar to others taken in rural parts of the US.[18] This suggests that the regional airshed over the Great Lakes is well mixed, and is representative of the overall tropospheric PCB concentrations at mid-latitudes. Sub-regional air concentrations are greater by factors of 3 to 5, such as those measured in Chicago.[30] Historical data indicate that current concentrations are less than half of what they were a decade ago (approximately 1 ng m^{-3}).[31,32]

Concentrations of PCBs in rain are also reasonably constant across the lakes, with more temporal variability observed at a given site than among sites. Concentrations for 1989–1991 ranged from 1 to 8 ng l^{-1}.[18]

Sediment Concentrations and Time Trends

Sediments provide current concentrations at the sediment–water interface, and when data are combined with the age of the sediment, they provide a complete historical record of concentration and accumulation rate with time. Thus a single sediment core from a depositional zone of a lake provides a half-century of information. The most recent sediment data are from a large study of several cores from lakes Superior, Michigan and Ontario which were dated and analysed for PCBs and 13 other contaminants.[33-35] The data were corrected for sediment focusing and differences in sedimentation rates. PCBs in sediment correlate well with the rate of PCB production and use, rising rapidly from the 1930s to a peak in the 1970s, followed by a decline to the present. The differences among lakes are consistent with the water data discussed above, with Lake Superior ($\sim$ 7 ng g^{-1} dry weight) having the lowest PCB levels both historically and currently, and lakes Ontario ($\sim$ 200 ng g^{-1}) and Michigan ($\sim$ 125 ng g^{-1}) having much greater levels, which reflect their higher loadings. The PCB accumulation rates will be discussed below.

[30] S. J. Eisenreich, Rutgers University, New Brunswick, NJ, unpublished data.

[31] D. L. Swackhamer, B. D. McVeety and R. A. Hites, *Environ. Sci. Technol.*, 1988, **22**, 664.

[32] J. B. Manchester-Neesvig and A. W. Andren, *Environ. Sci. Technol.*, 1989, **23**, 1138.

[33] S. J. Eisenreich, D. L. Swackhamer and D. T. Long, *Atmospheric Deposition of Toxic Contaminants to the Great Lakes: Assessment and Importance*, Great Lakes Protection Fund and USEPA, Chicago, IL, 1995, vol. 1.

[34] D. L. Swackhamer, S. J. Eisenreich and D. T. Long, *Atmospheric Deposition of Toxic Contaminants to the Great Lakes: Assessment and Importance*, Great Lakes Protection Fund and USEPA, Chicago, IL, 1995, vol. 2.

[35] D. T. Long, S. J. Eisenreich and D. L. Swackhamer, *Atmospheric Deposition of Toxic Contaminants to the Great Lakes: Assessment and Importance*, Great Lakes Protection Fund and USEPA, Chicago, IL, 1995, vol. 3.

3   PCB Fate and Transport

Processes of Fate and Transport

Concentrations are useful for assessing trends, but do not adequately describe the fate of a contaminant in the environment. The rates of movement of contaminants from one compartment to another are necessary to assess fate, and to construct models that can predict fate under different environmental conditions. Models that describe contaminant transport and fate can range from simple equilibrium box models to highly complex dynamic models. For modelling PCB fate, accuracy and precision are limited by our ability to describe the processes involved, and the availability of actual field data for calibrating and validating the models.

The processes that must be understood to evaluate transport and fate include the processes that bring PCBs into the lakes, processes which remove them and in-lake processes which may redistribute them.[36] The inputs include atmospheric deposition, gas absorption, tributary loadings, sediment resuspension and diffusion and direct discharges. Processes that remove PCBs from the water column include volatilization, sedimentation and surface water outflow. Exchange with groundwater is considered a negligible process in the Great Lakes. Loss of PCB due to degradation or photolysis is also considered to be negligible. While reductive dechlorination has been reported in a few isolated regions such as the Hudson River,[37] the high concentrations needed to stimulate such a process are found in only a few harbours in the Great Lakes (*e.g.* Waukegan Harbour[38]). Sediments in depositional zones of the lakes that have been examined for such a loss have not found any evidence for this process. The in-lake processes include the partitioning between dissolved and particulate phases, and transfer up the foodweb. The latter process will be considered separately in Section 4 below.

Some of the processes listed above are more important than others, and some are better understood than others. For models describing PCBs' fate in the Great Lakes, the major processes to be considered are atmospheric deposition processes, gas–water exchange, tributary loadings, sediment accumulation and resuspension.

Atmospheric Deposition

Deposition of PCBs from the atmosphere includes wet and dry depositional processes. Estimates of these processes can be obtained from measured air concentrations in the vapour and particulate phases which are applied to mass transfer equations, or they can be measured directly.

Wet deposition is the rate at which PCBs are removed by precipitation events, including rain, snow and fog. Precipitation removes vapour-phase PCBs by

[36] D. L. Swackhamer and S. J. Eisenreich, in *Organic Contaminants in the Environment*, ed. K. C. Jones, Elsevier, Amsterdam, 1991, p. 33.

[37] J. F. Brown, Jr., D. L. Bedard, M. J. Brennan, J. C. Carnahan, H. Feng and R. E. Wagner, *Science*, 1987, **236**, 709.

[38] J. B. Risatti, *Rates of Microbial Dechlorination of PCBs in Anaerobic Sediments from Waukegan Harbor*, Illinois State Geological Survey, Champaign-Urbana, IL, 1992.

partitioning of the chemical between the vapour and rainwater, and is controlled by Henry's law.[36] Precipitation also removes particulate-phase PCBs by in-cloud scavenging or by below-cloud washout. For PCBs, wet deposition of the particulate phase is greater than that for the vapour phase. For monitoring purposes, total wet deposition (sum of both vapour and particle scavenging) is measured directly by collecting and measuring precipitation over a known time frame. Such a precipitation network is part of the IADN programme, although variability in the data indicates that these measurements are subject to high uncertainty.[18]

The most recent estimates of wet deposition are from a synthesis of the IADN data along with all compiled research data current to 1992.[18] These estimates are shown in Table 2.

The estimates for wet deposition assume that snow and rain have the same scavenging efficiency, and are based on rain measurements. Only a few studies have examined snow deposition directly.[31,39] Earlier field measurements supported the assumption that snow scavenged particulate and vapour PCBs less efficiently than rain.[31] However, recent work has examined snow scavenging in greater detail, and concludes that snow is a more efficient scavenger than rain.[39]

Dry deposition is usually calculated from air particulate concentrations and given particle dry deposition velocities. Estimates based on this approach have concluded that dry deposition is a less important deposition mechanism than wet deposition by factors of 2 to 3.[40,41] However, a recent method reported by Holsen and workers for measuring dry deposition directly using mylar-coated plates indicates that dry deposition of PCBs may be more important than previously thought.[42] They have validated their technique by comparing measured deposition to that estimated from air measurements and particle size distributions. Dry deposition of very large particles ($>10 \mu m$) is a substantial part of the atmospheric flux, and becomes important near air particulate sources. These larger particles have short atmospheric residence times and become less important further away from particulate sources. This study demonstrated that dry deposition of PCBs was important to the southern part of Lake Michigan when winds blew from the south across Chicago, but the effect was not present in the northern part of the lake.[42]

Gas Exchange

The transfer of PCBs across the air–water interface is a combination of gas absorption and volatilization. The net effect of these two transfers is typically described by the two thin film model adapted by Liss and Slater,[43] as neither of these processes can be measured directly. This model is a function of the mass

[39] T. P. Franz, PhD Thesis, University of Minnesota, Minneapolis, MN, 1994.

[40] A. W. Andren, in *Physical Behavior of PCBs in the Great Lakes*, ed. D. Mackay, S. Paterson, S. J. Eisenreich and M. Simmons, Ann Arbor Publishers, Ann Arbor, MI, 1983, p. 12.

[41] S. J. Eisenreich, in *Sources and Fates of Aquatic Pollutants*, ed. R. A. Hites and S. J. Eisenreich, American Chemical Society, Washington, 1987, p. 393.

[42] T. Holsen, K. Noll, S. Liu and W. Lee, *Environ. Sci. Technol.*, 1991, **25**, 1075.

[43] P. W. Liss and P. G. Slater, *Nature*, 1974, **247**, 181.

Table 2 Current estimates of annual loadings and losses of PCBs to the Great Lakes (all fluxes are in kg yr^{-1})

	Superior	Michigan	Huron	Erie	Ontario
Wet deposition[a]	125	91	90	43	35
Dry deposition[a]	32	23	23	10	7.6
Tributaries	28[b]	650[b]	236[b]	741[b]	609[b]
	150[c]	375[e]			1335[f]
Net gas exchange (all lakes = volatilization)	1900[c,d]	5130[d]	2550[d]	1100[d]	700[d]
		680[e]			575[f]
Sedimentation	110[c]	1530[e]	646[d]	1076[d]	1900[g]
	240[d]	226[d]			396[d]
					968[f]
Outflow	60[c]	23[e]	–	77[h]	284[f]

[a]Ref. 18; 1989–1991.
[b]Ref. 22.
[c]Ref. 45; 1986.
[d]Ref. 18; 1987–1988.
[e]Ref. 21; 1991.
[f]Ref. 17; 1985.
[g]Ref. 68; 1991.
[h]Calculated from tabulated data and Niagara River flow rate of 1.75×10^{11} m^3 yr^{-1}.

transfer coefficients across the air-side and water-side films, the PCB concentration gradient between air and water, and the Henry's law coefficient. The air-side mass transfer coefficients are a function of wind speed, temperature and molecular diffusivities in air. The water-side mass transfer coefficients are a function of the Schmidt number, which in turn is estimated from the kinematic viscosity and molecular diffusivities in water, and temperature. Resistance through the water-side layer controls 70–80% of the mass transfer for PCBs.[27] Estimates of this flux are fraught with uncertainty due to a lack of paired over-water and water measurements taken over substantial time periods. Instantaneous fluxes have been calculated for paired air–water data obtained in fall 1991 and spring 1992 in lakes Michigan and Superior, respectively. The net instantaneous fluxes in Lake Michigan for September 1991 ranged from 24 to 220 ng m^{-2} d^{-1} out of the lake (net flux was volatilization).[28] The large variance was due to variations in wind speed and PCB air–water concentration gradients. Instantaneous net fluxes were calculated for Lake Superior in July 1988, August 1990 and May 1992.[27] These fluxes ranged from -40 to 110 ng m^{-2} d^{-1} (positive means volatilization), comparable to Lake Michigan.

Using the extensive IADN database for air vapour concentrations, temperature data obtained by NOAA satellite imaging, and assuming a constant dissolved water concentration with season,[44] seasonal fluxes and an annual PCB flux were calculated for both lakes (Figure 2).[27,28] These results are the most rigorous estimates of net gas-exchange of PCBs for any body of water currently available. The net annual flux for Lake Michigan was 12.3 μg m^{-2} yr^{-1}, with the greatest fluxes occurring in the fall. For a few individual days in the spring, the net flux was into the lake (absorption). However, the net seasonal flux for all seasons was out

---

[44] D. L. Swackhamer and A. Trowbridge, University of Minnesota, Minneapolis, MN, unpublished data.

Figure 2 PCB air–water fluxes over an annual cycle for Lake Superior (a) and Lake Michigan (b). Also shown are surface water temperatures. (Adapted from refs. 27 and 28)

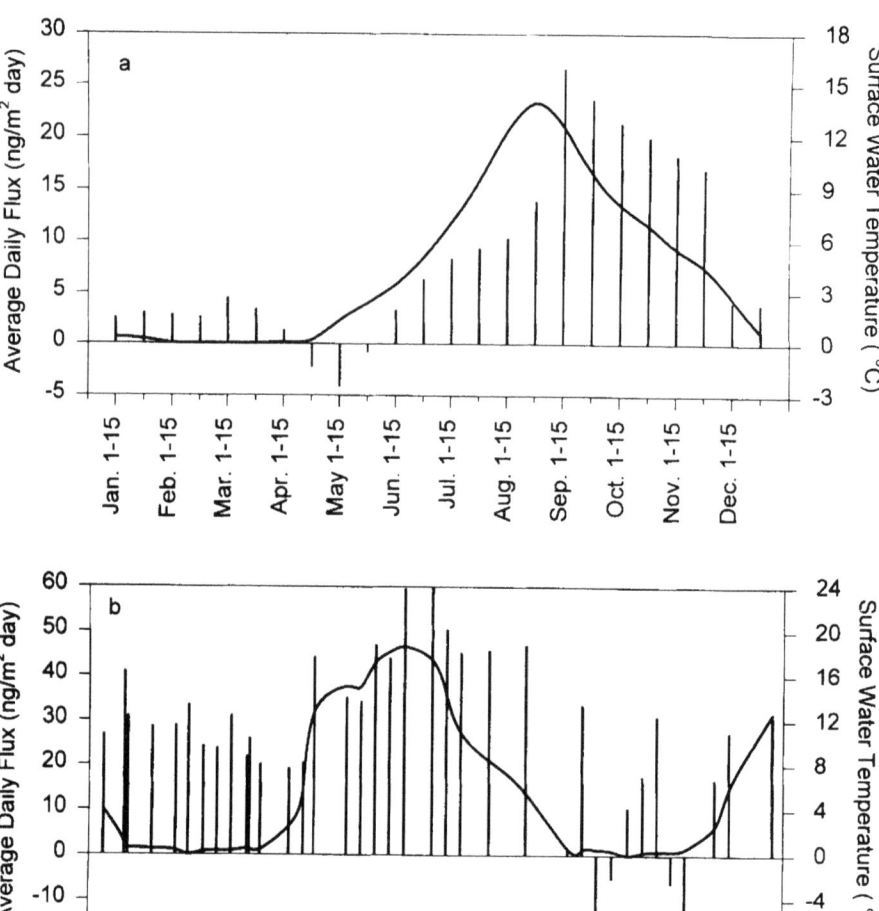

of the lake. The gross annual volatilization and absorption were 18.2 and 5.9 μg m^{-2} yr^{-1}, respectively. The net annual flux out of Lake Superior in 1992 was 3 μg m^{-2} yr^{-1}. Seasonal calculations revealed that vapour deposition into the lake occurs only briefly in the spring, while volatilization occurs throughout the rest of the year.

Estimates for gas exchange of PCBs for the other three lakes exist, but have not been made with the same level of rigour.[18] These estimates are given in Table 2. It is clear from these estimates that, on an annual basis, volatilization is more important than absorption for all of the Great Lakes, and that the lakes are a source of PCBs to the atmosphere as well as a sink.

Sediment Processes

Sediment core analyses have provided estimates of recent accumulations, as well as past accumulation records. PCB accumulation profiles from lakes Superior,

Figure 3 Accumulation of PCBs in sediments of lakes Superior, Michigan and Ontario with time[33]

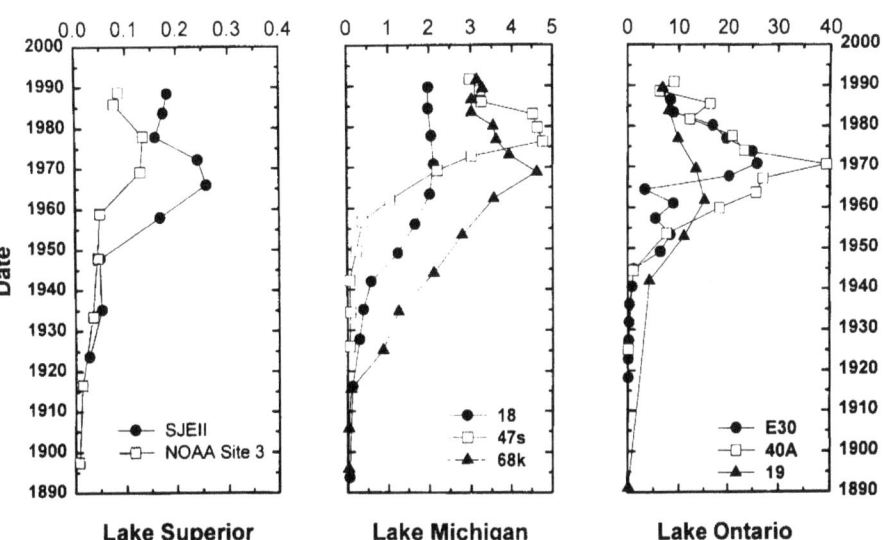

Total PCB accumulation (ng/cm² yr)

Lake Superior — SJEII, NOAA Site 3
Lake Michigan — 18, 47s, 68k
Lake Ontario — E30, 40A, 19

Michigan and Ontario are shown in Figure 3. Historically, accumulations peaked in the mid 1970s, which corresponds to the maximum production period in the US. Accumulations in Lake Superior[45] are similar to those from lakes receiving only atmospheric deposition,[16,31] supporting other evidence that non-atmospheric inputs of PCB to Lake Superior are negligible.[41] The accumulations and total areal burden of PCBs in Lake Michigan are about twice that of Lake Ontario, indicating that about half of the PCBs in Lake Michigan have come from non-atmospheric sources. The contribution from the connecting channels is a greater percentage of the input to the lower lakes, preventing comparisons of their sediment accumulations and burdens to the upper lakes.

Recent sediment accumulations are provided in Table 2. Current accumulations in Lake Superior are lowest, and Michigan and Ontario are highest. These loss rates can be directly compared to those from volatilization. For Lake Superior, net volatilization accounts for almost 90% of all losses.[20] For Lake Michigan, volatilization accounts for approximately 50% of all losses.[16,21] Estimates for the other lakes are 80%, 50% and 75% for lakes Huron, Erie and Ontario.[18]

Resuspension and diffusion processes may also play an important role in the transport and fate of PCBs in the Great Lakes. In fact, it has been modelled that transport of PCBs from sediments to the water column is the most significant input process in Lake Michigan.[46] Verification of this is extremely difficult, as it is one process that cannot be measured directly, nor are the physical mass transfer coefficients known with any certainty. It is probably the single largest source of

[45] J. Jeremiason, MS Thesis, University of Minnesota, Minneapolis, MN, 1993.
[46] D. Endicott and D. J. Kandt, *MICHTOX: A Mass Balance and Bioaccumulation Model for Lake Michigan*, USEPA, Large Lakes Research Station, Gross Ile, MI, 1992.

Figure 4 Mass budget of PCBs in Lake Superior for the year 1986. (Adapted from ref. 20)

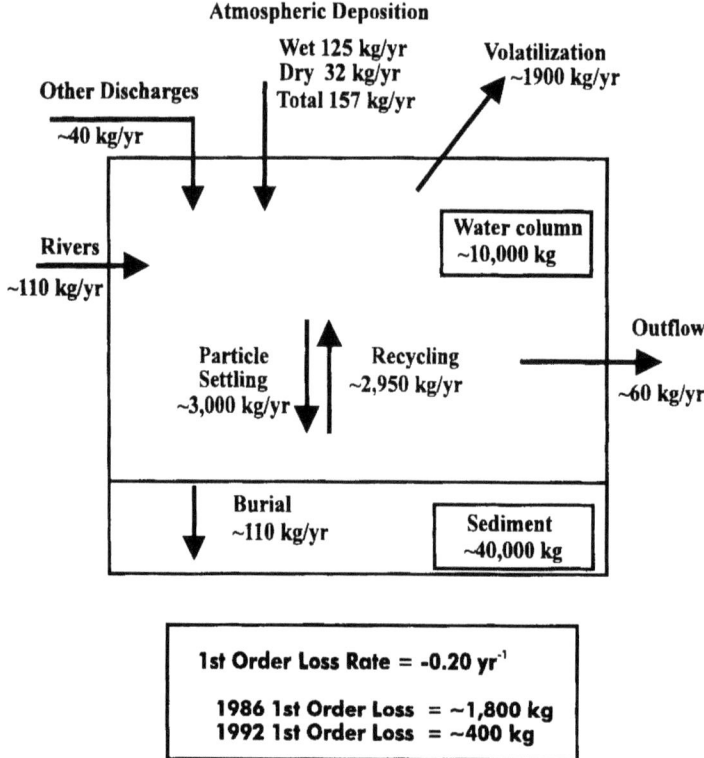

PCB Budget for Lake Superior, 1986

uncertainty in fate and transport models. No estimates of resuspension or diffusion are available for the other Great Lakes.

The estimates of the major input and loss processes can be combined in a mass budget to see if the system is near steady state, and to assess the relative importance of the different processes. This has been done for Lake Superior for the year 1986,[20] and Lake Michigan for the year 1991.[21] The mass budget for Lake Superior shown in Figure 4 underscores the importance of the atmosphere as both a major source and major sink of PCBs to and from the water column. It also illustrates the importance of PCB recycling at the sediment–water interface. While net burial rates are low, the gross flux to the sediments as measured by sediment traps can be very high. This recycling phenomena was first documented in Crystal Lake, Wisconsin[47] and then in the Great

[47] D. E. Armstrong, J. P. Hurley, D. L. Swackhamer and M. M. Shafer, in *Sources and Fates of Aquatic Pollutants*, ed. R. A. Hites and S. J. Eisenreich, American Chemical Society, Washington, 1987, p. 491.

Figure 5 Mass budget of
PCBs in Lake Michigan
for the year 1991.
(Adapted from ref. 21)

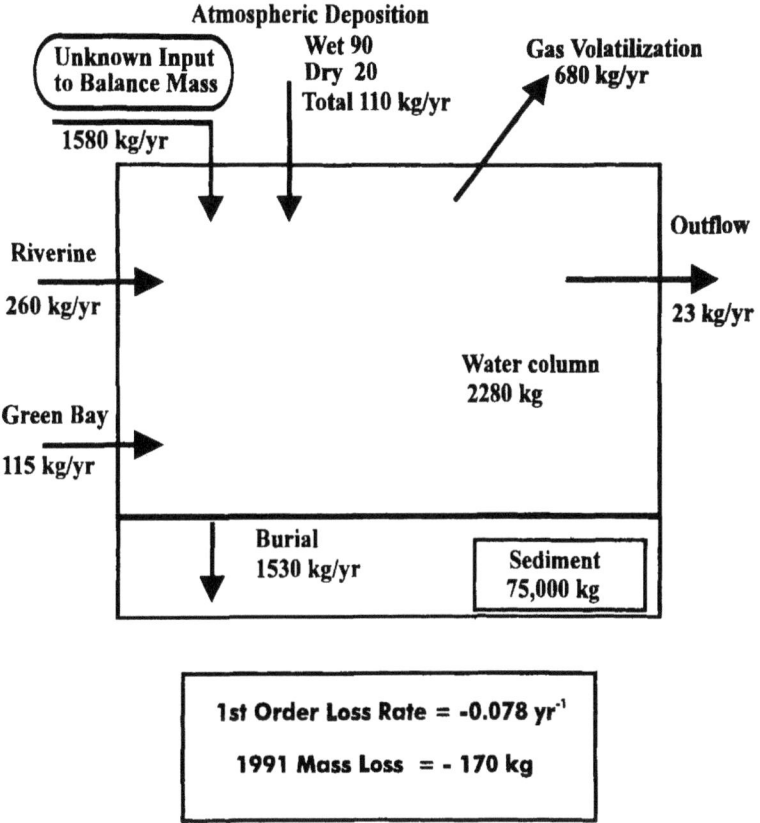

PCB Budget for Lake Michigan, 1991

Atmospheric Deposition

Unknown Input to Balance Mass

Wet 90
Dry 20
Total 110 kg/yr

Gas Volatilization
680 kg/yr

1580 kg/yr

Riverine
260 kg/yr

Outflow
23 kg/yr

Water column
2280 kg

Green Bay
115 kg/yr

Burial
1530 kg/yr

Sediment
75,000 kg

1st Order Loss Rate = -0.078 yr^{-1}

1991 Mass Loss = - 170 kg

Lakes.[48,49] The Lake Michigan budget (Figure 5) does not balance mass, indicating that the processes have not been fully described. The imbalance may be due to short-range dry deposition from the Chicago area, or to sediment-water exchange that is not adequately accounted for in the measurements.[21] The magnitude of the imbalance stresses the importance of these less-well understood processes.

Mass Balance Models

Mass budgets represent snapshots in time, constructed mostly with measurements. However, the processes themselves can be described quantitatively, and combined to describe contaminant fate and transport in mass balance models. The resource managers of the Great Lakes have promoted the use of mass balance models to be used as tools to help guide decisions and strategies for lake-wide management

[48] J. E. Baker, B. J. Eadie and S. J. Eisenreich, *Environ. Sci. Technol.*, 1990, **25**, 500.
[49] L. Lefkowitz, MS Thesis, University of Wisconsin, Madison, WI, 1989.

planning. In general, such models describe toxic fate as a result of changes in loadings due to a proposed regulatory decision, *i.e.* predict the change in sport fish concentrations of toxics as a result of dredging a contaminated area, or reducing a direct discharge, *etc.* The first effort in this regard was the Green Bay Mass Balance Study (GBMBS),[50] which developed a state-of-the-art dynamic toxics model concurrently with collection of an extensive database for calibrating and validating the model. A team of modellers linked a hydraulic model, a eutrophication model, a toxics fate model and a foodchain model such that known loadings of a contaminant would estimate water concentrations[51] and from there predict predator fish concentrations.[52] The field study was carefully tailored to the modelling effort to allow for precise calibration. The study focused on congener-specific PCBs in Green Bay, Lake Michigan. This successful five-year effort is now being expanded to a lake-wide model in the Lake Michigan Mass Balance Study. The collection of a calibration data set of PCBs in Lake Michigan is currently in progress.

Less sophisticated but highly useful models have been developed to assess the relative importance of different sources of toxics to the lakes. These screening level models can be equilibrium, steady state or dynamic models. Examples of these are the fugacity-based models of Mackay.[17]

4 Bioaccumulation and Foodweb Dynamics

Concentrations of PCBs in Great Lakes Biota

The dominant sport fish in the Great Lakes include the salmonines (lake trout, coho salmon, chinook salmon) at the top of the aquatic foodchain; important commercial species include whitefish, perch and bloater chubs. Economically, the several billion dollar sport fishery is of greater importance to the Great Lakes than the commercial fishery, particularly when the secondary benefits of tourism are considered. The lake trout must be stocked, and only self-reproduce in a few isolated areas of the Great Lakes. There is circumstantial evidence that the inability to reproduce is related to PCBs and other organochlorines.[53] The coho and chinook are non-indigenous, and are stocked. In conflict with the promotion of the sport fishery is the fact that the top predators have high body burdens of contaminants, particularly PCBs. Health consumption advisories exist for certain size classes of lake trout throughout the Great Lakes. These advisories suggest restrictions or bans on consumption for the general population, or for sensitive populations such as children or pregnant women, to prevent possible adverse health effects. Thus, on one hand fishery managers stock the lakes for fishing, and on the other hand health officials advise anglers to limit how much they eat.

[50] D. S. De Vault and H. J. Harris, *Green Bay Mass Balance Study Plan*, USEPA, Chicago, IL, 1987.
[51] J. V. DePinto, R. Raghunathan, P. Sierzenga, X. Zhang, V. J. Bierman, Jr., P. W. Rodgers and T. C. Young, *Recalibration of GBTOX: An Integrated Exposure Model for Toxic Chemicals in Green Bay, Lake Michigan*, Final Report to USEPA Large Lakes Research Station, Gross Ile, MI, 1993.
[52] J. P. Connolly, *Bioaccumulation Model for Green Bay Mass Balance Study*, Final Report to USEPA Large Lakes Research Station, Grosse Ile, MI, 1995.
[53] M. J. Mac, in *Toxic Contaminants and Ecosystem Health: A Great Lakes Focus*, ed. M. Evans, Wiley, New York, 1988, p. 389.

Concentrations of PCBs have been monitored in certain indicator species by both the USEPA and Environment Canada since the mid-1970s. The most extensive database is for lake trout, and these data are summarized in Figure 6. Current concentrations range from approximately $0.5 \mu g\ g^{-1}$ (wet weight in 60–70 cm whole fish) in Lake Superior to $1.5 \mu g\ g^{-1}$ in Lake Huron to $2 \mu g\ g^{-1}$ in Lake Ontario to $3 \mu g\ g^{-1}$ in Lake Michigan. The time trends in these data indicate that concentrations decreased dramatically from the mid-1970s to the mid-1980s, as expected in response to the restrictions and ban on PCB production. However, recent trends over the last 10 years indicate that the rate of decrease has slowed or stopped for all lakes.[22,54] Since water concentrations have decreased (discussed above), the lack of a similar decrease in fish is likely due to foodweb or diet preference changes. Lake trout in the Great Lakes receive the vast majority of their PCB burden from the foodweb compared to direct gill transfer from water.[55] This trend is even more pronounced in coho salmon.[22,56]

Contaminant trends are also monitored with herring gull eggs by Environment Canada. The observed trends are consistent with those in lake trout.[22] The lake trout integrate PCB trends throughout a lake; the gull eggs are useful for monitoring specific sites around a lake.[12]

There is a paucity of data for PCB concentrations in the lower trophic levels of the foodweb. There are a few isolated measurements from Lake Ontario plankton,[57] and an extensive database covering all seasons on phytoplankton, zooplankton and forage fish from Green Bay.[24,58] A similar database is currently being developed for all of Lake Michigan.[44] Preliminary data indicate that Lake Michigan phytoplankton and zooplankton have PCB concentrations of approximately 150 and $250\ ng\ g^{-1}$ dry weight, respectively. *Mysis relicta* have PCB concentrations of approximately $500\ ng\ g^{-1}$. There is clear evidence for biomagnification at each trophic level, as expected from the physical chemical properties of PCBs.

Foodweb Dynamics

Several foodweb models have been developed for the Great Lakes, driven by the concern for PCB contamination in sport fish.[59–60]

Efforts to model foodweb dynamics are hampered by calibration data, particularly the lower trophic levels, and by the large uncertainties associated with bioenergetic input parameters, such as PCB assimilation efficiencies for different organisms. Another major hindrance in aquatic foodweb models is a lack of understanding of the bioaccumulation of PCBs from water to the first trophic level, phytoplankton.[61,62] Research in this laboratory has revealed that

[54] D. S. De Vault, USEPA Great Lakes National Program Office, Chicago, IL, unpublished data.

[55] R. V. Thomann and J. P. Connolly, *Environ. Sci. Technol.*, 1984, **18**, 65.

[56] B. G. Oliver and A. J. Niimi, *Environ. Sci. Technol.*, 1988, **22**, 388.

[57] USEPA Great Lakes National Program Office, Chicago, IL, unpublished data.

[58] R. V. Thomann, J. P. Connolly and T. F. Parkerton, *Environ. Toxicol. Chem.*, 1992, **11**, 615.

[59] J. P. Connolly and C. J. Pedersen, *Environ. Sci. Technol.*, 1988, **22**, 99.

[60] F. A. P. C. Gobas, *Ecol. Model.*, 1993, **69**, 1.

[61] D. L. Swackhamer and R. S. Skoglund, in *Organic Substances and Sediments in Water*, ed. R. A. Baker, Lewis, Ann Arbor, MI, 1991.

[62] D. L. Swackhamer and R. S. Skoglund, *Environ. Toxicol. Chem.*, 1993, **12**, 831.

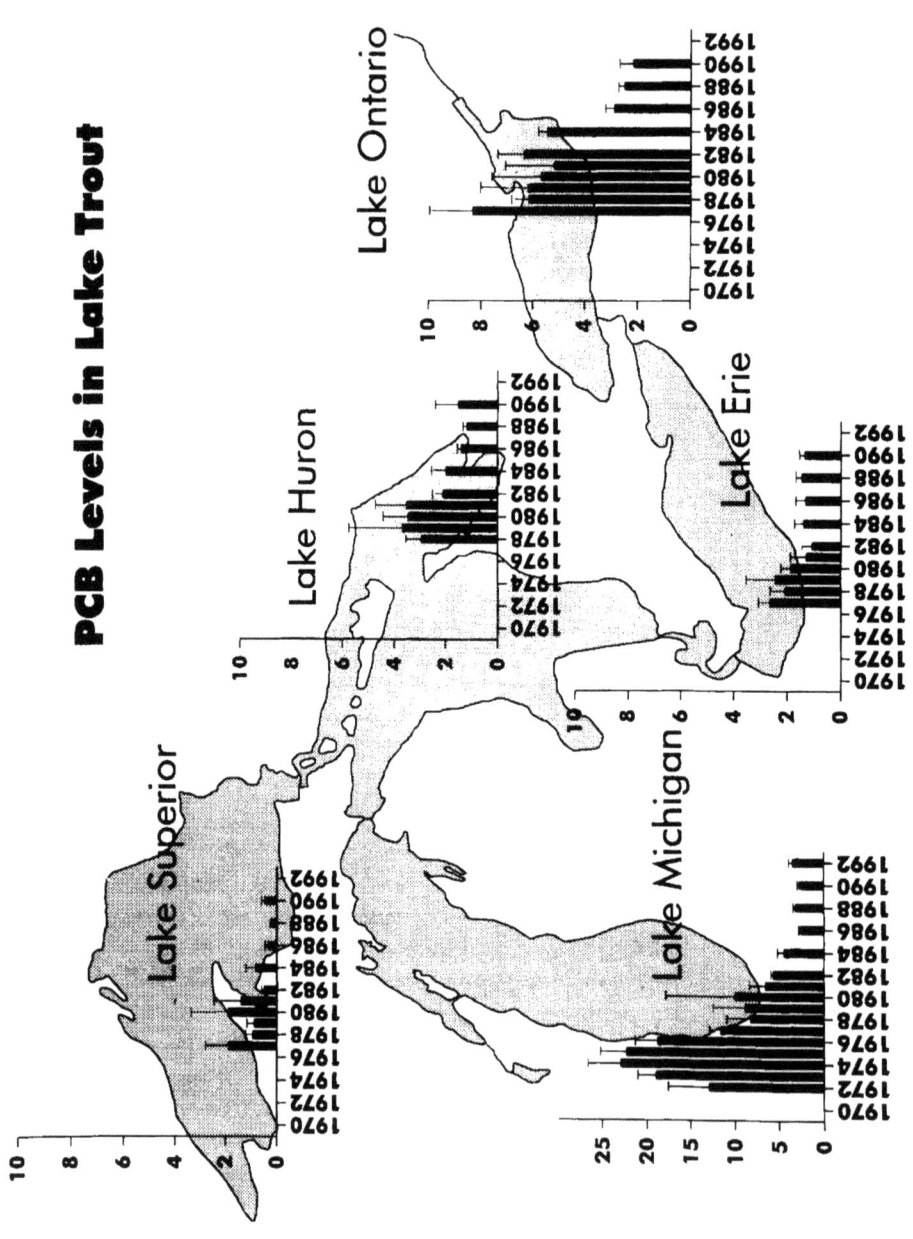

Figure 6 PCB concentrations (wet weight, whole body) in lake trout from the Great Lakes (walleye for Lake Erie) from 1971–1992. (Data from ref. 54; adapted from ref. 22)

phytoplankton uptake of PCBs is highly dependent on the growth rate of the phytoplankton,[62,63] and on the lipid content and type.[64] Phytoplankton bioaccumulation is poorly estimated from thermodynamics, and is best estimated using a kinetics model.[65,69] Improvements in estimating phytoplankton PCB uptake will reduce the uncertainty in foodweb models for estimating PCB concentrations in sport fish.

5 Summary and Conclusions

The Great Lakes have served as a focal point for PCB research. This research has provided an understanding of the processes controlling fate and transport of PCBs, and has led to the development of models than can be applied to other contaminants and water bodies. The processes of atmospheric deposition and net sediment accumulation are described adequately in these models, but the exchange at the sediment–water interface and seasonal depositional patterns need further improvements. While concentrations have declined in air, water and sediments over the last decade, trends in fish indicate a slowing or stopping of such a decline. Thus future research efforts should address the bioaccumulation process and foodweb dynamics, and the physical processes mentioned above.

6 Acknowledgements

The contributions of students Robert Skoglund, Annette Trowbridge and Roger Pearson, and the computer graphics by Joe Hallgren, are gratefully acknowledged. Original research described here has been funded over the years by the US EPA, the Minnesota Sea Grant Program of the National Oceanic and Atmospheric Administration, and the Great Lakes Protection Fund.

[63] R. S. Skoglund and D. L. Swackhamer, *Environmental Chemistry of Lakes and Reservoirs*, ed. L. A. Baker, American Chemical Society, Washington, 1993.

[64] K. Stange and D. L. Swackhamer, *Environ. Toxicol. Chem.*, 1993, **13**, 1849.

[65] R. S. Skoglund, K. Stange and D. L. Swackhamer, *Environ. Sci. Technol.*, 1996, **30**, 2113.

[66] J. Bicksler, MS Thesis, University of Minnesota, Minneapolis, MN, 1995.

[67] R. J. J. Stevens and M. A. Neilson, *J. Great Lakes Res.*, 1989, **15**, 377.

[68] C. Wong, D. L. Swackhamer, D. T. Long and S. J. Eisenreich, *Environ. Sci. Technol.*, 1995, **29**, 2661.

[69] K. C. Hornbuckle, State University at Buffalo, SUNY, Buffalo, NY, and S. J. Eisenreich, Rutgers University, New Brunswick, NJ, unpublished data.

Control of PCDD and PCDF Emissions from Waste Combustors

GEV H. EDULJEE AND PETER CAINS

1 Introduction

Previous contributions to this series[1-3] have highlighted the concern over emissions of polychlorinated dibenzo-*p*-dioxins (PCDDs) and polychlorinated dibenzofurans (PCDFs) to atmosphere from waste incineration processes. Regulatory control of these emissions has typically been addressed in two ways:

- By identifying what are believed to be the key design and operational parameters controlling emissions of PCDD/Fs, and stipulating suitable limits within which these parameters should be maintained.
- In addition to the above, imposing emission limits for PCDD/Fs in the stack gas.

For example, the UK's Chief Inspector's Guidance to Inspectors *Process Guidance Note IPR 5/3* requires new municipal solid waste incinerators to be designed to the following specifications:

- The combustion gases should be maintained at a temperature of at least 850 °C for at least 2 seconds in the presence of at least 6% oxygen.
- PCDD/F emissions should be below 1 ng m^{-3} (expressed as NATO-CCMS Toxic Equivalents, I-TEQ). The aim should be to achieve a guide emission concentration of 0.1 ng I-TEQ m^{-3}.

From an operational and regulatory perspective, the challenge is to develop control strategies which are effective in the field, but which are also practical to implement. For example, from a knowledge of the fundamental chemical and physical mechanisms of PCDD/F formation and their relationships to the prevailing physical and chemical conditions, it is possible to identify through laboratory experiments key operational parameters or surrogate emissions

[1] P. T. Williams, in *Waste Incineration and the Environment*, ed. R. E. Hester and R. M. Harrison, The Royal Society of Chemistry, Cambridge, 1994, p. 27.

[2] G. H. Eduljee, in *Waste Incineration and the Environment*, ed. R. E. Hester and R. M. Harrison, The Royal Society of Chemistry, Cambridge, 1994. p. 71.

[3] E. M. Steverson, in *Waste Incineration and the Environment*, ed. R. E. Hester and R. M. Harrison, The Royal Society of Chemistry, Cambridge, 1994, p. 113.

which can individually be correlated with PCDD/F emissions.[2,4] Control of these parameters would therefore imply control of PCDD/F emissions. However, in full-scale combustors, variations in feedstock composition and perturbations in operating conditions are often sufficient to mask correlations between measured PCDD/F emission levels and process parameters that tightly controlled bench scale experiments indicate are likely to be important. While some tests on full-scale plant have established correlations between, say, PCDD/F emissions and particulate emissions,[2,5] other studies have not found this to hold.[4,6] The lack of consistency in attempts to correlate measured PCDD/F levels globally against parameters identified as important in laboratory studies also indicates that simple, single, key operating parameters are not necessarily identifiable, and that combinations of factors determine PCDD/F formation and emission levels. This has important practical consequences, since it implies that a control strategy based on regulating simple process variables may not be amenable to a generalized prescription of preferred operating conditions.

This chapter commences with a brief review of the current basis for the development of control strategies relating to PCDD/F emissions from waste combustors. There follows a discussion of some of the plant trials undertaken to gain insights into the influence of operating variables on PCDD/F formation. The chapter ends with an analysis of potential control strategies for PCDD/F control in combustors, applying our current knowledge on the fundamentals of PCDD/F formation.

2 Early Investigations

PCDDs and PCDFs were first identified in incinerator stack emissions in the mid-1970s. Emissions of other trace organics (chlorobenzenes, chlorophenols, PAHs *etc.*) were also known to occur, but public and regulatory attention focused on the former compounds owing to their toxicity relative to other organic micropollutants, and highly publicized accidental releases from industrial installations.

Theoretical studies into the mechanisms of formation of PCDD/Fs in combustion systems, with the specific aim of elucidating emission control strategies, commenced in the early 1980s. Seminal contributions[7-9] examined free radical, homogeneous gas-phase reactions in the hot, combustion zone of municipal solid waste (MSW) incinerators, and concluded that this formation mechanism could not account for measured concentrations of PCDD/Fs in incinerator stack emissions. Measurements taken at various stages of the combustion and gas cleaning train of a MSW incinerator at Tsushima, Japan,

[4] P. W. Cains and P. H. Dyke, *Chemosphere*, 1994, **28**, 2101.

[5] D. D. Wallace, presented at the 1993 Incineration Conference, Knoxville, TN, 3–7 May, 1993.

[6] US EPA, *Combustion Emissions Technical Resource Document* (*CETRED*), Report No. EPA 530-R-94-014 (Draft), EPA, Washington, 1994.

[7] W. M. Shaub, *Technical Issues Concerned with PCDD and PCDF Formation and Destruction in MSW Fired Incinerators*, Report NBSIR 84-2975, National Bureau of Standards, Washington, 1984.

[8] W. M. Shaub, *Containment of Dioxin Emissions from Refuse Fired Thermal Processing Units – Propects and Technical Issues*, Report NMSIR 84-2872, National Bureau of Standards, Washington, 1984.

[9] W. M. Shaub and W. Tsang, *Environ. Sci. Technol.*, 1984, **17**, 721.

indicated enhanced PCDD/F concentrations as the combustion gases passed through the cooler, post-combustion zone. It was postulated that reactions between phenolic precursors (formed in the combustion zone as products of incomplete combustion, (PICs) and hydrochloric acid on the surface of flyash in the cooler (150–250 °C) parts of the incinerator were responsible for the observed elevated levels of PCDD/Fs in the stack gases.[10] Other research teams investigated heterogeneous surface-catalysed reactions operating in the temperature range 250–400 °C and confirmed the dominance of this formation mechanism.[11–14]

These studies suggested a mechanistic framework which could provide a basis for the development of PCDD/F emission control strategies:

- Incomplete combustion of organic wastes in the combustion chamber leads to the formation of solid carbon and organic fragments (PICs) which serve as organic precursors to the dioxin/dibenzofuran molecule.
- The waste provides a source of chlorine and of metals. The latter are incorporated into flyash, which carries over to the cooler (250–400 °C) post-combustion zone of the incineration system.
- The organic precursors are incorporated into, or adsorb onto the surface of the flyash in the post-combustion zone, and, following a complex sequence of reactions which are catalysed by metals (primarily copper) in the flyash, lead to the formation of PCDD/Fs along with other chlorinated trace organics.

From the early 1980s, pilot and full-scale incineration systems were studied to identify design and operating parameters which correlated with PCDD/F emissions.[15,16] Pre-1987 investigations emphasized combustion zone design and control, since the importance of post-combustion PCDD/F formation had yet to be fully appreciated. In keeping with the above mechanistic framework, parameters which related to efficient combustion of organic wastes (excess oxygen level, combustion temperature, CO level) were measured and correlated against PCDD/F emissions. However, the *ad hoc* and poorly controlled conditions under which measurements were generally taken limited the usefulness of these studies. The correlations claimed by some workers[16] were unconvincing when subjected to a rigorous statistical analysis.[10]

3 US and Canadian Studies

In the mid-1980s, systematic investigations to determine a set of generalized PCDD/F reduction/emission control strategies were commenced by the US

[10] B. Commoner, M. McNamara, K. Shapiro and T. Webster, *Environmental and Economic Analysis of Alternative Municipal Solid Waste Disposal Technologies: II. The Origins of Chlorinated Dioxins and Dibenzofurans Emitted by Incinerators that Burn Unseparated Municipal Solid Waste, and an Assessment of Methods of Controlling Them*, Center for the Biology of Natural Systems, Queens College, Flushing, New York, 1984.

[11] F. W. Karasek and L. C. Dickson, *Science*, 1987, **237**, 754.

[12] H. Hagenmaier, H. Brunner, R. Haag and M. Kraft, *Environ. Sci. Technol.*, 1987, **21**, 1085.

[13] H. Hagenmaier, M. Kraft, H. Brunner and R. Haag, *Environ. Sci. Technol.*, 1987, **21**, 1080.

[14] H. Vogg, M. Metzger and L. Stieglitz, *Waste Manage. Res.*, 1987, **5**, 285.

[15] E. Benfenati, F. Gizzi, R. Reginato, R. Fanelli, M. Lodi and R. Tagliaferri, *Chemosphere*, 1983, **12**, 1151.

[16] F. Hasselriis, presented at the ASME/IEEE Power Generation Conference, Portland, OR, October 19–23, 1986.

Environmental Protection Agency (US EPA) and by Environment Canada. These studies are described below.

US EPA Studies

Much of the early work on the control of organic emissions from combustors was conducted in the US as part of a programme involving the measurement of organic emissions and the formulation of control strategies. The concept of *Good Combustion Practice* (GCP) was introduced, the term being defined as 'those combustion conditions which lead to low emissions of trace organic pollutants'. Following a comprehensive study of three types of MSW incinerators (mass burn, RDF-fired and modular starved air units), the US EPA concluded in 1987 that low organic emissions could be achieved in properly designed and operated MSW incinerators, by a combination of good combustion control techniques and appropriate gas cleaning technology.

The rationale for the application of GCP to the control of organic emissions was that the latter were the products of incomplete combustion. Hence, optimization of combustion conditions to approach as closely as possible the theoretical ideal of complete combustion (*i.e.* combustion to CO_2, water, *etc.*), coupled with appropriate 'end-of-pipe' control strategies, should lead to reductions in trace organic emissions. The US EPA recommendations for GCP fell into three categories:

- Minimization of organic emissions to atmosphere through optimum design of the combustor.
- Operation of the combustor within its design specifications, with control systems to prevent excursions outside of the design envelope.
- Monitoring and verification of combustion performance, with continuous surveillance of key design and operating parameters.

The main design and operational parameters which required control in order to meet the above goals of GCP were identified as follows:

- Furnace temperature
- Underfire air capacity
- Overfire air requirement
- Excess air requirement
- Air distribution and mixing

Table 1 lists the recommendations for GCP to minimize organic emissions from combustors.[17]

The implementation of GCP will be discussed in Section 4 below.

Canadian Studies

In 1983, Environment Canada commenced a five-year National Incinerator Testing and Evaluation Program (NITEP) with the following objectives:[18]

[17] US EPA, *Assessment of Municipal Waste Combustor Emissions under the Clean Air Act*, Advance Notice of Rulemaking, 52 FR 25399, EPA, Washington, 7 July 1987.

[18] A. Finkelstein, presented at the US EPA/A&WMA Conference on Municipal Waste Combustion, Tampa, FL, April 16–19, 1991.

Table 1 US EPA recommendations for Good Combustion Practice (GCP) to minimize trace organic emissions from MSW, RDF and modular starved air combustors

Element	Components	Recommendations[a]	
Design	Temperature at fully mixed conditions	All:	1800 °F (980 °C) at fully mixed conditions
	Underfire air control	MB:	At least four separately adjustable plenums. One each under the drying and burnout zones and at least two separately adjustable plenums under the burning grate.
		RDF:	As required to provide uniform bed burning stoichiometry
		MSA:	No recommendations provided
	Overfire air capacity (not an operating requirement)	MB, RDF:	40% of total air
		MSA:	80% of total air
	Overfire air injector design	All:	That required for penetration and coverage of furnace cross section
Operation/ control	Excess air	MB, MSA:	6–12% oxygen in flue gas (dry basis)
		RDF:	3–9% oxygen in flue gas (dry basis)
	Turndown restrictions	All:	80–110% of design; lower limit may be extended by verification tests
	Start-up procedure	All:	On auxiliary fuel to design temperature
	Use of auxiliary fuel	All:	On prolonged high CO or low furnace temperature
Verification	Oxygen in flue gas	MB, MSA:	6–12% (dry basis)
		RDF:	3–9% (dry basis)
	CO in flue gas	All:	50 ppm on 4-hour average corrected to 12% CO_2
	Furnace temperature	All:	Minimum of 1800 °F (980°C) [mean] at fully mixed conditions
	Adequate air distribution	All:	Verification tests (adequately low exhaust emission of trace organics or combustion uniformity using in-furnace CO profiles)

[a]All = all combustors; MB = mass burn combustors; RDF = refuse-derived fuel combustors; MSA = modular-starved air combustors.

- To identify energy-from-waste technologies in Canada.
- To assess relationships among state-of-the-art designs, operations, energy benefits and emissions of organic and inorganic trace chemicals.
- To examine the effectiveness of emission control systems.
- To develop national guidelines for emissions from MSW incinerators.

The study predated the theoretical advances of the mid-1980s which identified post-combustion low-temperature PCDD/F formation reactions as a key consideration. The programme focused primarily on conventional combustion-related design and operational issues.

Mass-burn Technology Assessment. Extensive testing was undertaken on the Quebec Urban Community MSW incinerator in Quebec City, a mass-burn, water-wall combustor equipped with heat recovery boilers and a two-stage electrostatic precipitator. The objective was to determine the design and operational changes necessary to upgrade the incinerator to state-of-the-art, and to minimize emissions of organics and metals. Operating parameters such as MSW feed rate, excess air levels, distribution of underfire and overfire air, combustion temperature, *etc.*, were systematically varied, concurrent with measurements of CO, organics (total hydrocarbons, chlorobenzenes, chlorophenols, PCBs, PAHs, PCDDs and PCDFs), particulate levels, metals and acid gases. A one-sixth scale model of the furnace was constructed to test out possible design changes.

The trials were successful in that design and operating conditions were identified, which provided consistently lower emissions of organics and metals relative to both the original furnace design and operating regime, and to 'poor' operating conditions with the new design. Under 'good' operating conditions, emissions of each of the organic chemical groups were lowest; conversely, trace organic emissions were highest under conditions indicative of 'poor' operation (low furnace temperature, poor air distribution, high MSW feed rate, *etc.*). PCDD/F and particulate emissions were reduced by the same degree under the new design and operating conditions, suggesting a direct operational and mechanistic link.

Simple and multiple regression analysis provided relationships between organic and inorganic emissions and operating parameters. The most important of these were as follows:

- PCDD/F emissions were strongly correlated with primary gas flow, CO emissions, particulate emissions, copper emissions, chlorobenzene emissions and chlorophenol emissions. Primary airflow was the most influential operational setting for control of trace organic emissions.
- Particulate emissions correlated with CO emissions and primary gas flow.
- As a class, emissions of trace organics were minimized by a combination of control on CO, oxygen and furnace temperature. The best control models used three of the following four operational settings:
    - total airflow
    - primary/secondary air ratio
    - steam rate or refuse feed rate
    - secondary air front/rear ratio
- CO was the best single surrogate for the prediction of PCDD/F emissions. PAH and PCB emissions did not correlate well with CO emissions.
- The results of NITEP mirrored that of the 1987 US EPA recommendations for GCP (see Table 1). Using the modified design and control models identified in the study, 10- to 100-fold reductions in trace organic, particulate and metals emissions were achieved on a consistent basis.

Gas Cleaning Technology Assessment. In a second series of tests, two pilot-scale air pollution control systems were tested on the Quebec City MSW incinerator: a dry scrubber/fabric filter and a wet-dry (spray dryer) scrubber/fabric filter. Three sampling points were positioned on the pilot plant to determine gas characteristics and trace organics/metals composition at the inlet, pre-fabric filter and post-fabric filter. Operating parameters such as inlet temperature, pressure drop across the filter, particulate, CO and oxygen concentration, *etc.*, were measured.

For PCDD/F removal, overall removal efficiency across the scrubber unit was determined, masking any catalytic effect of trapped flyash. However, the tests did identify an increase in PCDD/F concentration in the flue gas with an increase in gas temperature, in the region 100–200°C.

4 Implementation of Good Combustion Practice

Introduction

From a regulatory standpoint, the most appropriate means of implementing GCP for minimization and control of trace organic emissions is to frame a set of general rules which can be applied to all combustors of a particular type. Five classes of criteria can be identified:

- Design criteria (for example, requiring a minimum gas-phase residence time of 2 seconds in the combustion zone).
- Operational criteria (for example, requiring a minimum furnace temperature of 850°C, or maintenance of a minimum excess oxygen level in the combustion gas).
- Measurement and control of surrogates such as CO and particulate emissions (for example, maintaining CO emission below $50\,\text{mg m}^{-3}$).
- Control regimes (restrictions of waste feeds; failsafe, interactive control systems; automatic shutdown procedures, *etc.*).
- Monitoring regimes (measurement and recording of combustion temperature; continuous monitoring of excess oxygen, CO emissions and combustion efficiency, *etc.*).

HMIP's Chief Inspector's Guidance Notes and European Union Directives on incineration provide examples of regulatory requirements and/or guidance which incorporate aspects of the above five types of criteria for GCP.

It has been noted that the imposition of specific GCP requirements (for example, maintaining a minimum furnace temperature, or maintaining CO levels of less than $50\,\text{mg m}^{-3}$) on small-scale commercial combustors has not been universally successful in ensuring low PCDD/F emissions.[19] This observation is in line with the comments in Section 1 above regarding the lack of consistency in regulating PCDD/F emissions through simple, single operating variables in isolation from other broader considerations.

In the sections below a few correlations suggested by the mechanistic model for PCDD/F formation are reviewed in the light of tests conducted on a wider range of pilot-scale and large-scale systems.

[19] J. D. Kilgroe, L. P. Nelson, PO. J. Schindler and W. S. Lanier, in *Incineration of Hazardous Waste – Toxic Combustion By-Products*, ed. W. R. Seeker and C. P. Koshland, Gordon and Breach, Philadelphia, 1992.

Particulate Emissions and PCDD/F Emissions

The NITEP study indicated a strong correlation between particulate emissions and PCDD/F emissions, suggesting that one emission control strategy would involve improving particulate collection efficiency in the gas cleaning train. In regulatory terms, this requirement of GCP would be achieved by specifying more stringent particulate emission limits. However, other studies have shown the absence of a correlation between particulate and PCDD/F emissions in small-scale combustion systems located in the UK.[4] A broader survey of the literature indicates more positive correlations, but also indicates the lack of a consistent pattern. Some recent studies are discussed below.

US Studies. In 1994 the US EPA published the results of an extended study of PCDD/F and particulate emissions from a variety of combustors fed with hazardous waste, in a report entitled *Combustion Emissions Technical Resource Document* (CETRED).[6] According to the report, US EPA 'is evaluating [particulate] emissions because controlling [particulate emissions] will control emissions of most toxic metals and toxic organic compounds absorbed into the [particulate matter]'.

The tests were conducted with the knowledge of the importance of post-combustion, low-temperature reactions, and represented an advance on the 1987 GCP study. The results of the trials on cement kilns, lightweight aggregate kilns, hazardous waste incinerators, liquid injection incinerators, fluidized bed incinerators, fixed hearth incinerators and hazardous waste boilers were summarized as follows:[6]

'While the [discussion on low-temperature reactions] emphasizes the role of particulate matter [PM] in the post-combustion production and capture of PCDD/PCDF, it is important to note that other parameters can dominate the production of PCDD/PCDF, and tight post-combustion control of PM emissions without attention to these other parameters can lead to higher, not lower, dioxin/furan emissions. It is also important to note that the above discussion does not imply that low PM emissions are necessarily accompanied by low dioxin emissions ... it is possible to minimize post-combustion formation of dioxin by limiting [air pollution control device] inlet temperatures (for example, by rapidly quenching combustion gases). When this occurs, low dioxin/furan emissions can be achieved without significant PM control ... it appears to be quite possible for [hazardous waste] burning facilities with moderate or even no PM control to exhibit low PCDD/PCDF emissions. Similarly, facilities can exhibit extremely high dioxin/furan emissions in the presence of very low PM emissions.'

The US EPA defined Best Current Operating Practice (BOP) as being the use of GCP combined with a temperature limitation of 350 °F (approximately 175 °C) on the inlet to post-combustion control devices. Rapid quenching of the combustion gases to below 175 °C was also regarded as BOP.

However, it does not follow that on an individual plant basis, consistent correlations between particulate emissions and PCDD/F emissions cannot be found. For example, an analysis of the US EPA test programme on clinical waste incinerators indicated the following.[5]

- PCDD/F emissions were strongly correlated with both particulate and CO emissions. In newer facilities with better combustion control, the strongest correlation was with CO, while in older facilities the strongest correlation was with particulate matter.
- There was evidence of interdependencies between individual operating parameters on a particular facility. For example, primary chamber temperature was also strongly correlated with PCDD/F emissions through its influence on CO and particulate emission levels.

These correlations have also been observed in other full-scale incinerator studies, for example in the NITEP programme.

Dutch Studies. In a study examining emissions from MSW incinerators operating in the Netherlands, a number of emission types were measured: acid, gases, organics, particulates, CO, *etc.*[20] The relationship between particulate and PCDD/F emissions was tested. In presenting a plot of these emissions, the report stated that 'no really clear correlations emerge.'

The lack of a correlation was also appreciated in a follow-up study in which the data presented in the Dutch report[20] (particulate concentration, water content, carbon monoxide concentration, furnace temperature, *etc.*) were compared with the PCDD/F emissions using statistical regression techniques.[21] The aim was to determine which operating and emission parameters were strongly related to PCDD/F emissions. The data set was divided into two groups on the basis of the type of air pollution control device installed:

- plants equipped solely with electrostatic precipitators (ESP);
- plants fitted with both an ESP and a wet scrubber (so-called 'Modern installation').

For the latter type of plant, it was found that the particulate concentration did not correlate with PCDD/F emissions. Only two parameters elicited statistically meaningful correlations: the furnace temperature and the carbon content of the filter dust collected in the ESP.

Negative observations of this nature should be treated with caution in view of the positive correlations obtained by other workers, and also because inter-plant comparisons would not be expected to elicit as consistent a correlation trend as for tests on a single plant.[4,5]

The Dutch study notes that:[20] 'In general, it can be concluded that low dioxin emissions ... only occur at low levels of CO, C_xH_y and particulates. The converse is not the case, however; low CO, C_xH_y and particulate levels by no means guarantee lower dioxin emissions.' This qualification is important: since low particulate emissions do not guarantee low PCDD/F emissions, it may not be cost-effective to insist on low particulate emissions on a particular plant if tests

[20] W. Slob, L. M. Troost, M. Krijgsman, J. de Koning and A. A. Sein, *Combustion of Municipal Solid Waste in the Netherlands*, Report No. 730501052, RIVM, Netherlands, 1993.

[21] G. Brem, R. Gort and L. B. M. van Kessel, *Theoretical and Experimental Modelling of Municipal Solid Waste Incineration*, TNO Report No. 92-269, TNO, Apeldoorn, Netherlands, 1992.

show that low PCDD/F emissions are being consistently achieved. However, in terms of establishing the general principles of an overall PCDD/F control strategy, requiring low particulate emissions would appear to constitute a sensible precautionary measure.

German Studies. Independent confirmation of the possibility of high particulate emissions being accompanied by low dioxin emissions is provided by the monitoring data collected on the TAMARA pilot MSW incineration plant at the Karlsruhe Nuclear Research Centre.[22] The pilot plant was fitted with cyclones and a wet scrubber. The particulate and total PCDD/F concentrations in Table 2 are downstream of the cyclones but prior to the wet scrubber. If the reported PCDD/F concentrations are converted to toxic equivalents, values in the region of 1 ng I-TEQ m^{-3} would be obtained. The data indicate that particulate emission concentrations far in excess of present-day regulatory requirements can be accompanied by low PCDD/F emissions.

A further discussion on the influence of particulate emissions on PCDD/F formation and emissions is provided in Section 5.

Operating Conditions and PCDD/F Emissions

Parametric studies on full-scale plant, examining the influence of operating parameters on PCDD/F formation, have been noted in Section 1 and Section 3. There is a remarkable consistency in the conclusions drawn in the studies on operating plants. Two studies[23,24] are discussed for illustrative purposes.

Parametric tests conducted by US EPA on the Montgomery County South MSW incinerator have been reported.[23] Three tests at each of six different test conditions were conducted to evaluate the effects on PCDD/F emissions of electrostatic precipitator (ESP) inlet temperature, sorbent injection into the furnace and into the duct leading to the ESP, and combustion temperature. The key findings were as follows:

- Poor combustion conditions in the furnace and high ESP temperatures resulted in the highest PCDD/F formation rates as measured in collected flyash.
- Formation of PCDD/Fs within the ESP was strongly dependent on the ESP operating temperature. A reduction in operating temperature from 300 °C to 200 °C decreased PCDD/F stack emissions by a factor of 10.
- PCDD/F concentrations increased across the ESP, even at a low ESP operating temperature of 149 °C.
- Greater removal efficiency of particulates also correlated with lower stack emissions of PCDD/Fs.

Other factors not explicitly accounted for in the tests may also have influenced the findings. For example, introduction of the sorbent into the ESP would have increased the particulate loading in the ESP, potentially increasing the surface

[22] H. Vogg, H. Hunsinger and L. Stieglitz, *Chem. Eng. Technol.*, 1990, **13**, 221.

[23] J. D. Kilgroe, W. S. Lanier and T. R. von Alten, presented at the US EPA/A&WMA Conference on Municipal Waste Combustion, Tampa, FL, April 16–19, 1991.

[24] J. Jager, M. Wilken, A. Beyer, H. Rakel, B. Zeschmar-Lahl and E. Jager, *Chemosphere*, 1993, **27**, 141.

Test number	Particulates (mg m^{-3})	PCDD/Fs (ng m^{-3})
T19/1	500	78.5
T19/4	953	103
T19/7	1032	<50
T19/10	947	<32

Table 2 Particulate and total PCDD/F concentrations at the TAMARA pilot MSW incineration plant

area available for PCDD/F formation. A lower ESP temperature would have lowered the gas velocity through the unit (therefore resulting in higher particulate removal), but would also have resulted in a lower PCDD/F formation rate.

The second study[24] describes tests undertaken at four different MSW incinerators under a variety of operating conditions reflecting both good and poor combustion practice. The latter was investigated by measuring operating parameters and emissions during 'cold blowing', corresponding to a low production of steam, a low combustion temperature and a high CO release. An interesting aspect of the tests is that the boiler cleaning operation known as 'soot blowing' was also investigated. Over the period of soot blowing, a 10-fold increase in particulate emissions was observed.

The outcome of the tests was similar to that of study on the Montgomery County South incinerator, and are summarized below:

- During periods of 'cold blowing' arising from the spontaneous reaction of the primary air supply system to a blockage in the feed hopper, an increase in primary air flow led to a fall in combustion temperature, an increase in oxygen concentration and a three-fold increase in PCDD/F emission concentrations. Utilizing supplementary support fuel to maintain constant combustion conditions during periods of cold blowing resulted in a sharp decrease in PCDD/F emissions.

- Cooling the combustion gases post-boiler from 470°C to 265°C resulted in a sharp decrease in PCDD/F emissions. While this is attributed largely to increased adsorption of PCDD/Fs to particulate matter at the lower temperature, material which is subsequently removed from the gas phase by the pollution control device, it is also possible that the reduction is a consequence of the shift to the lower end of the window of PCDD/F formation of 250–400°C.

- During soot blowing, a 10-fold increase in particulate emission concentrations was accompanied by a 30-fold increase in PCDD/F emission concentrations prior to the pollution abatement device. Accounting for particulate removal in the fabric filter, an overall three-fold increase in PCDD/F emission concentrations was observed during periods of soot blowing.

Based on the tests, a number of means by which PCDD/F formation could be minimised were suggested.[24] These are listed in Table 3.

Comparisons with the GCP requirements listed in Table 1 and the discussion relating to the NITEP trials indicates an overall consistency of approach to the control of PCDD/F formation in combustors. The efficacy of controlling the halogen content of the waste fuel is examined in Section 5.

Recommendations concerning the design of MSW incinerators
- Avoid the risk of cold zones in the combustion chamber by providing:
  –air flow regulation according to demand
  –means for equal distribution of MSW on the combustion grate
  –means for homogenization of input MSW
- Avoid 'cold blowing' effects through preheating of combustion air
- Install continuously performing technologies for cleaning boiler surfaces
- Provide for a cooling section for flue gas to enhance adsorption of gaseous PCDD/Fs onto particulates
- Consider a high-performance flue gas purification and abatement system
- Reduce flue gas temperature after the boiler system below 250°C
- Reduce flue gas temperature at the ESP considerably below 250°C

Recommendations concerning operating conditions
- Keep operation conditions constant
- Avoid CO peaks and high oxygen surplus
- Reduce particulate and soot deposition through more frequent cleaning
- Reduce halogen availability in combustion chamber through addition of halogen fixing additives such as limestones and amines
- Increase the transformation of gaseous PCDD/Fs into particulate-bound form by the addition of adsorptive solids such as activated carbon
- Improve general flue gas purification through the addition of limestone and activated carbon

Recommendations concerning waste input
- Reduce halogen availability in the combustion chamber by excluding certain materials from the MSW input
- Exclude brominated compounds (for example, electronic scrap) from the MSW
- Exclude PVC from the MSW in order to reduce halogen input

Source: Jager *et al.*[24]

Precursor Emissions versus PCDD/F Emissions

A corollary of the precursor-mediated route to PCDD/F formation is that a reduction in precursor concentrations in gases exiting the combustion chamber will have a beneficial effect in reducing the propensity for PCDD/F formation in downstream equipment. Organic precursors such as chlorobenzenes and chlorophenols can be formed as a result of gas-phase reactions in the primary and secondary combustion chambers of incinerators. These precursors to PCDDs and PCDFs adsorb onto flyash in the cooler parts of the incinerator such as the boiler and gas cleaning units such as ESPs, and are then subjected to catalytic reaction in the presence of copper to form PCDDs and PCDFs. The greater the quantity of precursors formed in the furnace, the greater is the propensity for dioxin formation in downstream equipment (assuming downstream conditions are such that dioxins can be formed).

The formation of precursors is dependent on the temperature at which the waste is combusted (and, in addition, residence time and mixing characteristics), and hence the degree of destruction of the organic constituents in the primary and secondary chambers. Higher combustion temperatures equate to a higher degree of destruction and hence to lower concentrations of potential precursors leaving the furnace zone. Since destruction of potential organic precursors is dependent on the combustion conditions in the furnace, one might also expect interactions between combustion temperature and precursor emissions, combustion air control (distribution and oxygen concentration) and precursor concentrations, and possibly ultimate correlations between some or all of these parameters and PCDD/F emissions.

A recent study has examined potential correlations between emissions of organic precursors and PCDD/F emissions.[25] A large number of studies on full-scale operational plants have observed a proportional relationship between PCDD/F emissions and emissions of other trace organics such as chlorophenols and chlorobenzenes. An example of predictions based on such correlation is shown in Figure 1.

It is important to note that the correlation does not necessarily imply a causal link between the presence of a precursor and the presence of PCDD/Fs in the stack gases: the two types of substances may have been formed by different and unrelated reaction pathways. However, the presence of a correlation does imply that changes in the combustion process and hence in the underlying reaction schemes affect both types of emissions to an equal extent.

A number of factors can obscure or destroy a proportional relationship between precursor and PCDD/F emissions concentrations. For example, the mixing characteristics of the combustion chamber are equally important, and represent a key design variable. Poor mixing of wastes and combustion air could negate the advantage gained by operating at a higher temperature (and hence a higher propensity for precursor destruction). Another confounding factor is the relationship between precursor concentrations and the presence of particulate matter. The reactions leading to the formation of PCDDs and PCDFs are surface catalysed. Hence their formation will be dependent on the number of active sites on the surface of the flyash, and therefore on particulate composition/morphology and particulate concentration. However, to a large extent the formation of precursors and the generation of particulates are independent processes. It is possible to envisage a case where poor combustion results in an excess of organic precursors but (owing to low ash content) low particulate concentrations relative to the available precursors. The resulting PCDD/F concentrations will also be low, but it is likely that an increase in particulate concentration will be accompanied by an increase in dioxin formation, up to the point where the number of active sites and the number of precursor molecules are in balance. Beyond this point, a increase in particulate concentration will not result in an increase in dioxin formation.

Conversely, it is possible to generate very low concentrations of organic precursors (for example, because of a high furnace temperature) but relatively high concentrations of particulates owing to a high ash content of the waste. Since the number of active reaction sites exceeds the number of precursor molecules, a further increase in particulate concentration will not be accompanied by an increase in dioxin formation. This may explain the data presented in Table 2 for

---

[25] T. Öberg and J. Bergström, *Organohalogen Compd.*, 1992, **8**, 197.

Figure 1 Observed and predicted PCDD/F emissions from a MSW incinerator, using chlorobenzenes and chlorophenols as surrogates. (Taken from Öberg and Bergström[25])

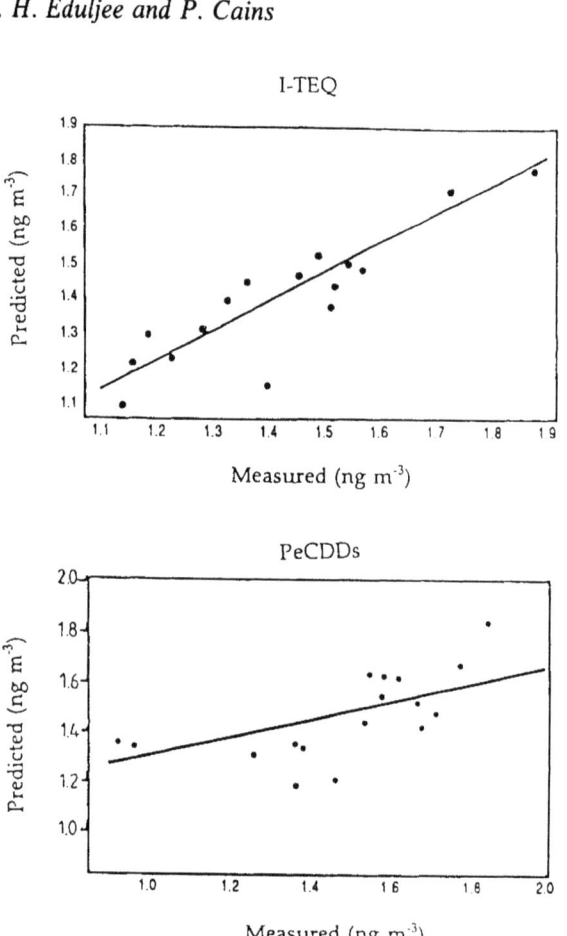

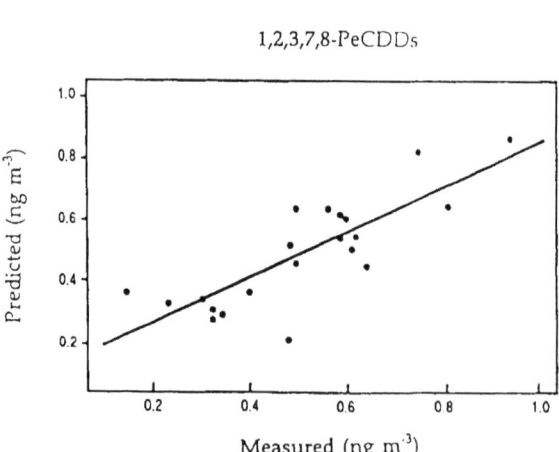

the TAMARA facility, Tests T19/1 and T19/7, in which a two-fold increase in particulate concentration ($500 \, \text{mg m}^{-3}$ to $1032 \, \text{mg m}^{-3}$) is not accompanied by a corresponding increase in PCDD/F concentrations ($78.5 \, \text{ng m}^{-3}$ to $< 50 \, \text{ng m}^{-3}$).

Another variable which will affect the correlation is particle size. Since the reactions of interest are surface catalysed, the quantity of PCDD/Fs formed should strictly be proportional to the available surface area, which in turn will depend on particle size. A greater proportion of small particles in a sample of flyash will result in a greater surface area being available for reaction than another sample of the same weight containing a greater preponderance of larger particles. Inspection of the data on Dutch MSW incinerators[20] shows that there is a wide variation in particle size distribution between the various incinerators: 'Zaanstad B' has 14% of the flyash below $0.95 \, \mu\text{m}$ and 50% above $20 \, \mu\text{m}$, whereas 'Rosendaal' has 40% of flyash below $0.95 \, \mu\text{m}$ and 10% above $20 \, \mu\text{m}$. Therefore, two identical particulate concentrations from two separate incinerators could well present different surface conditions for the reactions of interest.

Thus, precursor concentrations and their effect on PCDD/F formation cannot be divorced from upstream particulate concentrations.

5 Reaction Fundamentals and Control Strategies

General Principles

The principles of good combustion practice (GCP) in Section 3 can be restated in a manner which better relates to the mechanistic aspects of PCDD/F formation and to field operational and control regimes. Updating GCP to include catalysed heterogeneous reactions in the post-combustion zone, the following goals of GCP can be identified:

- Maximize the destruction of organics in the combustion chamber, so as to prevent the carryover of uncombusted organics or products of incomplete combustion (PICs) into the post-combustion zone. This in turn will reduce the likelihood of PCDD/F formation in the post-formation zone.
- Minimize particulate carryover into the post-combustion zone. Since the dominant PCDD/F formation reactions have been identified as being catalysed by the surface of flyash, reduced carryover should lower the likelihood of PCDD/F formation.
- Minimize the potential for low-temperature catalysed reactions in the post-combustion zone by minimizing the time the gases and particulate matter spend in the temperature region 250–400 °C or by suppressing the catalytic activity of flyash.
- Minimize emissions of PCDD/Fs by employing end-of-pipe control strategies.

In terms of developing control strategies relating to the control of PCDD/F formation and emissions, the above goals can be discussed under four headings:

- Control of feedstock
- Control of the combustion process
- Control of the post-combustion process
- End-of-pipe strategies

The combined control of combustion and post-combustion processes constitutes the equivalent of US EPA's Best Current Operating Practice (BOP) discussed above. In this section, we draw out relevant observations concerning the fundamentals of PCDD/F formation in order to inform the development of potential control strategies.

Control of Feedstock

Three issues are discussed under this heading: (1) the effect that different feed materials might have on the propensity for PCDD/F formation in combustors; (2) whether restrictions in feedstock composition are likely to have a beneficial effect in terms of reduced PCDD/F emissions; (3) whether the manner in which the feedstock is presented to the combustor influences PCDD/F formation. Each issue is discussed below.

Effect of Different Feed Materials. The question as to whether different waste types have intrinsically different compositional characteristics which impact on the propensity for PCDD/F formation during combustion has not been specifically addressed in this study. The fact that all waste types, and indeed all fuels, including fossil fuels such as coal, generate PCDD/Fs on combustion suggests a common reaction framework which applies to all combustors fired with any type of fuel or waste type. However, studies have indicated that flyash generated from MSW combustion and from the combustion of chemical waste differ in their catalytic activity, the former being more active per gramme of material.[26] This either reflects the compositional variations of different starting materials, or different combustion conditions under which these wastes are treated. For example, it is possible that the breakdown of lignin structures within wastes fed to a MSW incinerator results in a more active flyash than would be generated in a chemical waste incinerator. This issue has yet to be resolved.

In terms of PCDD/F control strategies, an important point is that combustors should be specifically designed to accept the waste in question, so as to ensure that the requirements of GCP and BOP are met regardless of the type of waste. Thus, a designer of a clinical waste incinerator should allow for the fact that medical waste invariably arrives at the facility in sealed bags or containers, that the composition of the waste could vary markedly from bag to bag, and that there is relatively little scope for inspection and equalization of loads. The design of the waste reception and loading equipment, feed regime, and of the combustor will therefore be influenced by these considerations. In the case of the combustion of wood chips, other factors may be more important and result in a different design of the combustion system. The key point is that if GCP and BOP are observed, then the issue of waste type is of secondary importance.

Restrictions in Feedstock Composition. There has been considerable debate as to whether removal of chlorine-containing components of MSW (such as PVC) prior to combustion contributes to a lowering of PCDD/F emissions relative to a conventional mass burn operation (see Table 3). The rationale for this suggestion

[26] K. P. Naikwadi, I. Albricht and F. W. Karasek, *Chemosphere*, 1993, **27**, 335.

is that chlorination reactions at post-combustion temperatures cannot proceed to completion and thus result in maximum yields of PCDD/Fs if HCl delivery to the reaction sites is reduced. In order to achieve the latter, the principal sources of chlorine in the feed material should be removed prior to combustion.

PCDD/F formation in combustors is a side-reaction which, in terms of percentage yield, is inconsequential relative to the dominant oxidative reactions between organic matter and chlorine in MSW and oxygen in combustion air. Literature sources indicate that the quantity of HCl in process gases is at best a secondary determinant in influencing PCDD/F yields, and is much less important than the temperature–time window.[27,28]

PCDD/Fs are produced in trace quantities, and the demand for HCl participating in the chlorination reactions is correspondingly very small. While removal of materials such as PVC could potentially result in a significant reduction in the total amount of HCl generated by the combustion of MSW in the furnace, this will not necessarily impact on the small quantity required for PCDD/F formation. In other words, there is sufficient chlorine present in the remaining MSW, clinical waste or other waste-based feedstock after removal of PVC to satisfy the requirements of the PCDD/F formation reactions, even under optimum formation conditions. PVC/plastics removal is therefore unlikely to affect emissions of PCDD/Fs, all other operational conditions remaining constant.

Recent trials on laboratory, pilot and full-scale plant have tended to confirm the lack of a beneficial effect on PCDD/F emissions, when chlorine-containing components of MSW and other waste types are withdrawn from the feedstock to a combustor.[29,30] As a strategy for controlling PCDD/F formation, our view is that removal of chlorine-containing materials such as PVC is unlikely to prove effective.

Similar arguments should apply to the likely effectiveness of strategies based on reducing levels of metals and potential organic precursors in the waste stream, at least with regard to influencing PCDD/F emissions. Removal or control of metals or sources of metals such as mercury and cadmium may still be necessary in order to achieve the required emission concentrations for these chemicals.

Presentation of the Feedstock.   Mention was made above of the need to consider the manner in which the waste is presented to the combustor. The aim is to ensure that the requirements of GCP are met, once the waste has been introduced into the combustor.

Combustion is best controlled when the waste is homogeneous, both in physical and compositional terms.[2] Thus, during industrial waste combustion this can be achieved by blending the incoming waste streams to a consistent, controlled composition prior to combustion. Solid material can be shredded

[27] I. Fängmark, B. van Bavel, S. Marklund, B. Strömberg, N. Berge and C. Rappe, *Environ. Sci. Technol.*, 1993, **27**, 1602.

[28] B. K. Gullett, P. M. Lemieux and J. E. Dunn, *Environ. Sci. Technol.*, 1994, **28**, 109.

[29] American Society of Mechanical Engineers, *The Relationship between Chlorine in Waste Streams and Dioxin Emissions from Combustors*, Final Draft Report, ASME, Washington, 1995.

[30] Association of Plastics Manufacturers in Europe, *MSW Combustion – Effects of Mixed Plastics Waste addition on Solid Residues and Chlorinated Organic Compounds*, APME, Brussels, 1994.

prior to being fed into the incinerator in order to reduce particle size, and ensure increased contact with combustion air, and uniform burning. A graphic illustration of the effect of waste preparation on combustion control has been presented elsewhere.[2] Feeding large items of waste into an incinerator resulted in large excursions in CO levels, with spikes rising to as high as 2000 ppm in the stack gas owing to the difficulty in controlling both the mixing with combustion air and the combustion temperature. Shredding the waste to a smaller and more uniform size ensured better control of combustion conditions, and resulted in a lower and smoothed CO release.

Different types of waste will require different handling techniques. For example, clinical waste is invariably sealed at source, and shredding of this material prior to it being fed into the combustor would not be considered good practice. The principle, that the combustor and associated handling requirements should reflect the type of waste to be treated, remains an important consideration for the control of PCDD/F emissions.

Control of the Combustion Process

Conventional control strategies dealing with the combustion process are well summarized by the US EPA's requirements for good combustion practice (GCP) listed in Table 1, and by the requirements listed in Table 3. There are three issues to be considered: (1) whether current knowledge of the fundamentals of PCDD/F formation provides either a justification for these requirements, or represents a conflict with presently held views on PCDD/F control; (2) whether current knowledge suggests alternative, less costly, means of control; (3) whether some or all of these requirements need to be modified and/or supplemented with new requirements in the light of more recent knowledge.

Relevance of Current Thinking to Good Combustion Practice. With respect to issues (1) and (2), current thinking on the mechanistic aspects of PCDD/F formation does not contradict the requirements for GCP as set out in Section 5 and by the US EPA in 1987 (see Table 1), nor does it suggest that any of these requirements are superfluous. Precursor concentrations, temperature regimes, *etc.*, are directly impacted by operational practices which have been demonstrated in both bench-scale and full-scale tests either to reinforce or to negate the goals set out for GCP in a predictable and reproducible manner. This issue is discussed further in Section 6.

As with feedstock preparation, uniform mixing of the waste with combustion air and adequate turbulence in the combustion chamber are important to ensure consistent and controlled combustion conditions. In order to implement GCP on old plant, it may be necessary to examine and redesign, for example, the primary air distribution system and air control regime, and introduce baffles to improve turbulence. Such modifications have been shown to be effective in improving combustion control, and consequently in reducing PCDD/F emissions.[18] These are not novel control strategies, nor are they necessarily inexpensive to implement, but current thinking relating to GCP does reinforce the need for greater customization in incinerator design. The use of computational fluid dynamic (CFD) modelling of incinerator plants is a means of simulating the material and energy distribution patterns in an incinerator, and is an excellent tool for visualizing the flow patterns in an incinerator, hence aiding in the design of plant configurations and control systems.[31,32]

PCDD/F Formation Mechanisms.   With regard to issue (3), a recent development is the evidence of in-flight formation of PCDD/Fs. Measured formation rates for in-flight formation have been much higher than those for static formation,[28,33,34] but they inevitably apply over relatively short residence times of gas–solid contact. Moreover, it has been proposed that the enhanced catalytic activity is associated with freshly produced ash, and that this activity diminishes inherently over periods of the order of 0.1–1 second.[34]

However, in-flight mechanisms may determine the ultimate PCDD/F emission levels that are achievable in well designed and operated plant, since it is very difficult to envisage a design that does not involve passing combustion gas with some particulate loading through the temperature window of 250–400 °C. The application of in-flight calculations could assist in estimating potential limiting formation levels under various post-combustion scenarios.[28] Since the in-flight formation rate is in units of ng (total PCDD/F) g^{-1} (flyash) min^{-1}, it follows that the current calculation method would prescribe low particulate loading and short residence time in the temperature window 250–400 °C as the measures most likely to minimize PCDD/F formation.

Another potential limiting factor relating both to the feed composition and the combustion process is the incomplete destruction of PCDD/Fs present in the incoming waste. For example, if the MSW contained a trace quantity of PCDD/Fs at a concentration of, say, 50 μg I-TEQ tonne^{-1} (wet) and if 99% was destroyed in the combustor, then the uncombusted PCDD/Fs would contribute a concentration of 0.1 ng I-TEQ N m^{-3} in the exit gas, assuming that physical and chemical processes in the post-combustion zone (for example, adsorption on activated carbon) do not alter the composition of the combustion gases during their passage to the stack. This process is not well understood, and requires further elucidation in bench-scale tests.

Control of the Post-combustion Process

Several issues need to be considered in relation to control of the post-combustion stage of a combustor:

- Control of the temperature window in which optimum PCDD/F formation occurs: 250–400 °C.
- Minimization of particulate loading.
- Minimization of the time particulate matter spends in the post-combustion zone.
- Inhibition of the catalytic activity on flyash surface.

Each issue is discussed in turn.

[31] V. Nasserzadeh, J. Swithenbank, D. Scott and B. Jones, *Waste Management*, 1991, **11**, 249.
[32] V. Nasserzadeh, J. Swithenbank, D. Lawrence and N. P. Garrod, *Trans. Inst. Chem. Eng.*, 1995, **73B**, 212.
[33] E. Altwicker, *Sci. Total Environ.*, 1991, **104**, 47.
[34] R. Konduri and E. Altwicker, *Chemosphere*, 1994, **28**, 23.

Control of the Temperature Window for PCDD/F Formation. Bench-scale and full-scale trials are unequivocal in identifying the post-combustion temperature as a key operation variable influencing the formation of PCDDs and PCDFs. There is now a general consensus that maintenance of post-combustion conditions in the gas cleaning system below about 200°C is desirable.

There will always exist intermediate temperatures in sections of the post-combustion train between the furnace temperature (~ 900°C) and the conditioner/ESP/fabric filter temperature ($\sim 200\%$°C), notably in the boiler and economizer sections, and it therefore follows that PCDD/F formation cannot be entirely suppressed. However, good operating practice as currently recommended by equipment suppliers centres on the need to minimise buildup of particulate matter on equipment surfaces subjected to temperatures within the formation range so that the residence time of particulate matter subjected to these temperatures is minimized (see below).

Minimization of Particulate Loading. Our current appreciation of static and in-flight formation of PCDD/Fs indicates that if all other operational parameters were kept constant, lower particulate concentrations in the post-combustion zone would result in lower PCDD/F formation. Particulate removal prior to the gases entering the post-combustion zone is one control measure that merits examination. Conventional technology has concentrated exclusively on end-of-pipe particulate removal systems (ESPs, fabric filters, carbon filters, *etc.*), but with the recent development of filtration media capable of operating at high temperatures, the potential exists for the removal of a significant proportion of the particulate matter emanating from the furnace prior to its entry into the boiler and pollution abatement system. Drawing on static and in-flight PCDD/F formation mechanisms, a fall in particulate concentration in the furnace off-gases should be paralleled by a fall in PCDD/F production, assuming that all other requirements of GCP are observed. A recent trial on a fullscale clinical waste incinerator has demonstrated the efficacy of this control technique.[35]

Minimizing the Residence Time of Particulate Matter. The operation of boilers, economizers, ESPs and other post-combustion equipment at around 3200°C is generally regarded as conducive to the production of high PCDD/F emissions. High formation rates in boilers, ESPs, *etc.*, appear to occur primarily due to high static formation on trapped particulates and, in the case of ESPs, electrostatic enhancement of PCDD/F levels in the gas phase.[36] It follows that an important aspect of controlling emissions is the prevention of particulate build-up on plant and pipework operated in the temperature region associated with formation. This should be taken together with the minimization of all gas–particulate contact in the formation temperature window. However, the requirement to cool combustion gases implies that such conditions, and hence some PCDD/F formation, cannot be eliminated entirely.

[35] A. Halász, *Waste Manage. Res.*, 1996, **14**, 3.
[36] R. Kolluri and E. Altwicker, *Haz. Waste Haz. Mater.*, 1994, **11**, 145; M. Tanaka, I. Watanabe, M. Hiraoka, Y. Takizawa, Y. Masuda, R. Takeshita and K. Yagome, *Chemosphere*, 1989, **18**, 321.

Inhibition of Catalytic Activity. The fact that the dominant (*i.e.* heterogeneous) PCDD/F formation reaction is surface-catalysed offers a means of reducing PCDD/F formation in the post-combustion zone of a combustor. The addition of ammonia to the flue gases suppressed the formation of PCDD/Fs on flyash surfaces.[37] Since ammonia is also used as a reactant in the reduction of nitrogen oxides (NO_x) to nitrogen, the simultaneous control of NO_x and PCDD/F emissions using conventional de-NO_x technology has been suggested.[38,39] The results of successful pilot scale trials using ammonia injection technology and descriptions of commercial systems have been reported.[40,41]

Another approach to the suppression of catalytic activity has been taken by workers at the University of Waterloo.[26,42–44] After conducting extensive laboratory trials on flyash obtained from a variety of combustion plants, an amine-based 'destroyer/inhibitor' mixture was formulated. This reactant was injected into the boiler of a MSW incinerator in an amount that represented 7–10% of the flyash loading in the flue gas; the 'destroyer' in the temperature window 590 + 50°C, and the 'inhibitor' in the temperature window 375 + 50°C. Overall reductions in PCDD/F formation of 80–94% were claimed. This approach has yet to be made commercially available.

End-of-Pipe Strategies

These include the control of particulate emissions, adsorption of PCDD/Fs by activated carbon or other substrates, and decomposition of PCDD/Fs or catalytic destruction.

Control of Particulate Emissions. As noted above, PCDD/F formation is via a surface-catalysed mechanism. It may therefore be thought that capture and removal of the particulate matter would necessarily result in a corresponding reduction in PCDD/F emission concentrations. If this logic were applied as a control strategy in isolation, it would suggest progressive lowering of particulate emissions as a simple means of PCDD/F emission control. Yet as a large number of parametric studies have shown, it does not follow that a correlation between PCDD/F emissions and particulate emissions is necessarily observed in full-scale incinerators under all operating conditions. The reaction sequence is complex, and a number of variables affect the rate of formation and quantities of PCDD/Fs produced, for example:

[37] H. Vogg, M. Metzger and L. Stieglitz, *Waste Manage. Res.*, 1987, **5**, 285.
[38] V. Boscak and G. Kotynek, presented at the US EPA/A&WMA Conference on Municipal Waste Combustion, Tampa, FL, April 16–19, 1991.
[39] H. Hagenmaier, K. H. Tichaczek, H. Brunner and G. Mittelbach, presented at the US EPA/A&WMA Conference on Municipal Waste Combustion, Tampa, FL, April 16–19, 1991.
[40] L. Takacs and G. L. Moilanen, presented at the US EPA/A&WMA Conference on Municipal Waste Combustion, Tampa, FL, April 16–19, 1991.
[41] Lurgi Energie und Umwelt GmbH, International Symposium Separation and Treatment of Pollutants from Waste Incineration Flue Gases, Frankfurt, September 1993.
[42] K. P. Naikwadi and F. W. Karasek, *Chemosphere*, 1989, **18**, 1219.
[43] K. P. Naikwadi and F. W. Karasek, *Organohalogen Compd.*, 1994, **19**, 383.
[44] F. W. Karasek and K. P. Naikwadi, *Organohalogen Compd.*, 1994, **19**, 315.

- The type of organic precursors present in the combustion gases, and the concentrations of these precursors.
- The availability of the flyash surface, since other chemicals present in the gases compete for a limited number of adsorption sites, and reactions other than those resulting in PCDD/F formation also occur.
- The reactivity of the flyash surface.
- The amount of carbon and of metals (especially of copper) present in the flyash.
- The residence time of the flyash particles in the temperature zone of interest – whether they are held up on boiler tubes or ESP plates, or whether they pass rapidly through as in the case of a wet scrubber.

Following formation, PCDD/Fs will partition between the solid and the gas phases depending on the temperature and the nature of the solid surface. The efficacy of particulate capture as a means of control of PCDD/F emissions therefore also presupposes conditions under which PCDD/Fs are primarily associated with the solid phase.

For these reasons, particulate removal will not necessarily result in a proportional reduction in PCDD/F emissions. PCDD/F emissions can be controlled by means other than limiting particulate emissions (*e.g.* by controlling PCDD/F formation).

Adsorptive Processes. The use of activated carbon, sprayed into a dry/semi dry scrubbing unit along with lime or less frequently packed in an adsorption unit positioned after the particulate removal device and prior to the stack, has become a standard component in gas cleaning trains as a means of PCDD/F control on all sizes of plant fed with MSW or clinical waste. Other adsorptive media such as zeolites are also being tested. The inclusion of an adsorptive device in combustion systems fired with wood and agricultural wastes is not normally contemplated, and as noted above, an interesting issue to be resolved is whether different waste types generate flyash of different activities relative to PCDD/F formation.

Catalytic Destruction. It has been reported that catalysts employed for the selective catalytic reduction (SCR) of NO_x emissions also demonstrate the ability to decompose organohalogen compounds, including PCDD/Fs.[39] Successful pilot trials at MSW and hazardous waste incineration facilities indicated that PCDD/F emission concentrations of <0.1 ng I-TEQ m^{-3} could be achieved in the absence of ammonia when titanium dioxide-based SCR catalysts were maintained between 200–350°C.

6 Summary

Our analysis of PCDD/F formation mechanisms and results from parametric trials on bench, pilot and full scale plant tends to reinforce rather than supplant existing strategies for reduction and control of PCDD/F emissions. Mechanistic considerations supply an underlying rationale for the requirements of Good Combustion Practice, many components of which were formulated before the reaction pathways were elucidated in laboratory experiments. The key to the

R^2	Variables in model			
	CO (corrected)	NO_x (SDI)	H_2O (SDI)	Furnace temperature
0.79	X			
0.82	X	X		
0.89	X	X	X	
0.93	X	X	X	X

Source: Environment Canada, 1994.

R^2	Variables in model			
	RDF Moisture	Rear wall over fire wall	Undergate air flow	Total air
0.31	X			
0.39	X	X		
0.59	X	X	X	
0.67	X	X	X	X

Source: Environment Canada, 1994.

implementation of GCP has been amply demonstrated in the bench scale parametric studies and in the full scale NITEP trials: because of the inter-dependency of the operating variables and their interaction in terms of the effect on combustion conditions, it is likely that control of only a selection of operational parameters to the exclusion of others will not provide an overall optimum in terms of minimizing PCDD/F formation. All relevant operational parameters need to be controlled in concert in order to achieve the BOP goals outlined in Section 5.

Recent tests provide an excellent example of the control of PCDD/F emissions from MSW combustion facilities.[46] Tables 4 and 5 reproduce the results of a multiple regression analysis on operating variables relevant to the combustion system (*i.e.* before the combustion gases enter the pollution abatement equipment).

In the so-called predictive model illustrated in Table 4, progressively better correlations with PCDD concentrations in flue gas exiting the combustor are obtained as 'monitoring variables' comprising the concentrations of CO, NO_x and water and the furnace temperature are successively combined into a single overall control model. When all four monitoring variables are combined, excellent prediction of PCDD concentrations is obtained. However, when 'control variables' such as waste moisture content, rear wall air flow, total overfire air flow, and underfire air flow are correlated with PCDD emission concentrations, the overall fit is much less effective (see Table 5). A similar trend was observed

[45] M. Hiraoka and S. Okajima, *Organohalogen Compd.*, 1994, **19**, 275.

[46] Environment Canada, *National Incinerator Testing and Evaluation Program: The Environmental Characterization of Refuse-derived Fuel (RDF) Combustion Technology*, Draft Report No. EPS 3/UP/7, Environment Canada, Ottawa, 1994.

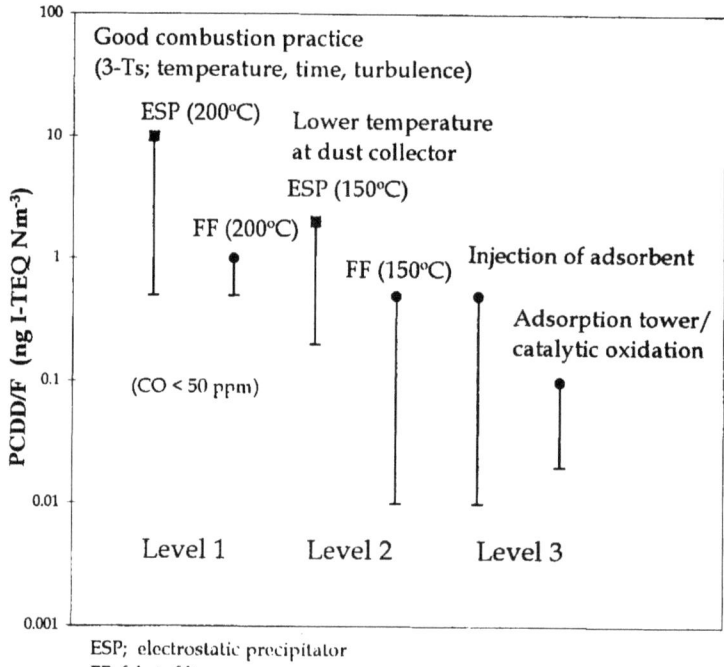

Figure 2 Control of PCDD/Fs in MSW incinerators. (Taken from Hiraoka and Okajima[45])

between plant variables and PCDF emissions.

The improvement in predictive power when 'monitoring variables' are combined points to the desirability of optimizing a range of operating variables in concert. Optimising furnace temperature also requires optimization of air flow, which in turn impacts on CO concentrations and NO_x production. Lack of control on any single operational variable is likely to negate any reduction in PCDD/F formation. The excellent correlation coefficient of $R = 0.93$ in Table 4 also indicates that optimization of all four monitoring variables will result in consistently low PCDD/F emissions from the furnace.

The poor correlation between PCDD/F emissions and 'control variables' ($R^2 = 0.31$–0.67) suggests that the mixing and flow characteristics of the furnace play an important role in ensuring optimum combustion conditions, and that it is not sufficient merely to monitor furnace input variables.[31,32] For example, it is possible to supply an adequate quantity of combustion air to the furnace, but intermittent blockages on the grate, the lack of turbulence in sections of the furnace or a sluggish air supply control system could result in less than optimum combustion conditions, and therefore higher PCDD/F emissions. It is essential that PCDD/F control strategies are underpinned by interactive monitoring of furnace conditions (and by extension, post-combustion operating conditions) if consistently low PCDD/F emissions are to be achieved and maintained.

The implementation of GCP, coupled with control of post-combustion conditions, can affect PCDD/F emissions as summarised in Figure 2 for an MSW incinerator.[45]

GCP alone will permit consistent achievement of PCDD/F emissions of 0.5–10 ng I-TEQ m^{-3}, as confirmed by the NITEP trials on the modified Quebec Urban Community MSW incinerator. Maintenance of GCP and operation of the pollution abatement equipment below 200 °C will help achieve further reductions to 0.1 ng I-TEQ m^{-3} and below. Catalytic oxidation or treatment with activated carbon enables the emission range to be maintained below 0.1 ng I-TEQ m^{-3} with greater consistency.

7 Acknowledgements

The authors are grateful to the Department of Trade and Industry, through the Energy Technology Support Unit (ETSU), for funding this work and granting permission for publication. The views expressed in this chapter are those of the authors and do not necessarily represent those of ETSU or the Department of Trade and Industry. The authors especially wish to thank Mr Patrick Dyke of ETSU for the valuable assistance and advice he has offered throughout the study.

Subject Index